87 Topics in Current Chemistry

Fortschritte der Chemischen Forschung

Micelles

Springer-Verlag Berlin Heidelberg GmbH 1980

This series presents critical reviews of the present position and future trends in modern chemical research. It is addressed to all research and industrial chemists who wish to keep abreast of advances in their subject.

As a rule, contributions are specially commissioned. The editors and publishers will, however, always be pleased to receive suggestions and supplementary information. Papers are accepted for "Topics in Current Chemistry" in English.

ISBN 978-3-662-15406-9   ISBN 978-3-540-34748-4 (eBook)
DOI 10.1007/978-3-540-34748-4

Library of Congress Cataloging in Publication Data. Main entry under title: Micelles.
(Topics in current chemistry; 87) "Author index, volumes 26–87." Bibliography: p.
Contents: Lindman, B. and Wennerström, H. Micelles. Amphiphile aggregation in aqueous solution.
– Eicke, H.-F. Surfactants in nonpolar solvents. Aggregation and micellization. 1. Micelles – Addresses,
essays, lectures. I. Series. QD1.F58 vol. 87 [QD549]. 540'.8s. 79-26530. [541'.34514]

© by Springer-Verlag Berlin Heidelberg 1980
Originally published by Springer-Verlag Berlin Heidelberg New York in 1980
Softcover reprint of the hardcover 1st edition 1980

Contents

Micelles
Amphiphile Aggregation in Aqueous Solution

Björn Lindman and Håkan Wennerström

Divisions of Physical Chemistry 1 and 2, Chemical Center, University of Lund, P.O.B. 740, S-220 07 Lund, Sweden

Table of Contents

1 Introduction

The many remarkable physico-chemical properties of aqueous surfactant systems, as well as their numerous practical applications, can be referred to the tendency of the nonpolar groups to avoid contact with water at the same time as the polar part tends to be strongly hydrated. The adsorption of surfactants at interfaces between aqueous solutions and air, another liquid phase or a solid is one consequence of this, the extensive aggregation into various types of large aggregates termed *micelles* — from lat. *micella* meaning small bit — and liquid crystalline phases is another. The aggregation into micelles is strongly cooperative and it has similarities to a phase separation.

Amphiphilic substances, of which surfactants are examples, have with one *hydrophilic* part and one *hydrophobic* part a marked spatial separation of their solvation properties. Shown in Fig. 1.1 are formulae and space-filling molecular models of a variety of typical amphiphilic substances which display extensive aggregation in the presence of water. It is found that not only the presence of distinct hydrophobic and hydrophilic parts but also other features, such as steric factors, are decisive for the aggregation process. Extensive cooperative aggregation is found for amphiphiles with a single long straight alkyl chain, while compounds with bulky nonpolar groups, like dipalmitoyl lecithin and Aerosol OT (cf. Fig. 1.1), do not display association into micelles in aqueous solution. A typical property of such compounds is instead the ability to take up considerable amounts of water in the formation of liquid crystalline phases in which there is a partitioning into extensive hydrophilic and hydrophobic regions. On the basis of this behavior, such compounds are often referred to as *swelling amphiphiles* in contrast to typical micelle-forming surfactants which are said said to be *nonswelling*. There is no sharp distinction between the two types, and conditions may change appreciably with temperature. Another useful classification of the amphiphiles refers to the polar group, i.e., one distinguishes between *ionic*, which may be cationic or anionic, and *nonionic*, which may be zwitterionic or dipolar, amphiphiles.

This treatment concerns amphiphile association in aqueous solution and will attempt to give an account of our present understanding of several aspects of micelle formation. In the succeeding section a phenomenological description of aqueous amphiphile solutions will be presented; there is included an enumeration of physico-chemical properties as well as a discussion of how features of the chemical structure influence aggregation. The relation to amphiphile aggregates in other phases will also be briefly considered in Sect. 2. Section 3, accounting for the thermodynamics of micelle formation, discusses what forces are responsible for the peculiar cooperative aggregation. The *hydrophobic effect,* which is of a wide interest in connection with the formation of many biological structures, is considered in some detail. Section 4 contains a broad outline of the structure of micellar solutions. Topics treated include micelle size and shape, hydrocarbon chain conformation, counterion binding, hydration and solubilization. Dynamic and kinetic aspects of micelles are treated in Sect. 5, for example the rate of monomer exchange, the rate of micelle formation and dissociation, solubilization kinetics, micelle mobility, and intramicellar motions. It is becom-

3

a. $CH_3(CH_2)_{11}OSO_3^- Na^+$

b. $CH_3(CH_2)_7 CO_2^- K^+$

c. $CH_3(CH_2)_{15} \overset{+}{N}(CH_3)_3 Br^-$

d. $\langle\rangle \overset{+}{N}(CH_2)_{11} CH_3 Cl^-$

e. $CH_3(CH_2)_{11} (OCH_2CH_2)_8OH$

f. $CH_3(CH_2)_{11} \overset{+}{N}(CH_3)_2 CH_2 COO^-$

g. $CH_3(CH_2)_{11} \overset{\downarrow}{\underset{O}{N}(CH_3)_2}$

h. $CH_3(CH_2)_3 \underset{\underset{C_2H_5}{|}}{CH} CH_2O\ CO\ CH_2$

$CH_3(CH_2)_3 \underset{\underset{C_2H_5}{|}}{CH} CH_2O\ CO\ CHSO_3^- Na^+$

i. $CH_2OCO(CH_2)_{14} CH_3$
$CHOCO(CH_2)_{14} CH_3$
$CH_2O\overset{O^-}{\underset{\underset{O}{||}}{P}}O(CH_2)_2 \overset{+}{N}(CH_3)_3$

j. (steroid structure with OH groups and $CH(CH_3)CH_2 CH_2 COO^- Na^+$)

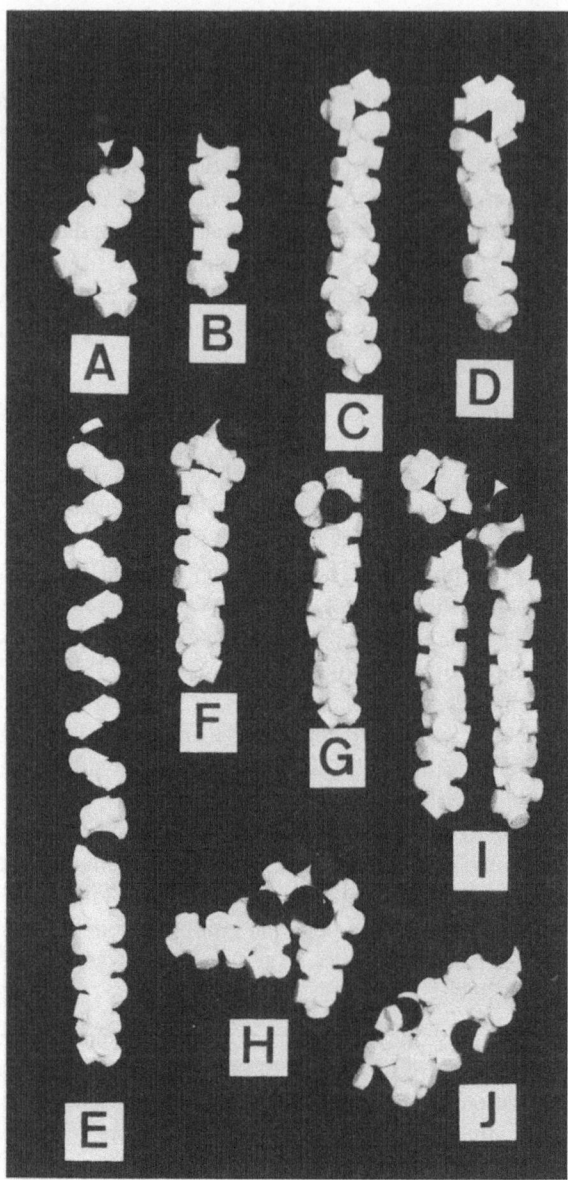

Fig. 1.1 Amphiphilic compounds. Chemical formulae (*left*) and molecular models (*right*) of the following compounds:

a. sodium n-dodecylsulfate (SDS)
b. potassium n-octanoate
c. n-hexadecyltrimethylammonium bromide (CTAB)
d. n-dodecylpyridinium chloride
e. n-dodecyloctaethyleneglycolmonoether
f. N-n-dodecyl-N,N-dimethylbetaine
g. dimethyl-n-dodecylamineoxide
h. sodium di-2-ethylhexylsulfosuccinate (Aerosol OT)
i. dipalmitoyllecithin
j. sodium cholate

Of these *a* to *g* are typical micelle-forming substances. In general the all-trans conformation of the alkyl chains is shown but in reality there is an appreciable gauche population. This is indicated for the case of SDS. The molecular models have been built by Thomas Anderson on the basis of the novel system designed by B. R. Thomas (B. R. Thomas "Molecular models for research and teaching in organic and biological chemistry – a model system designed for easy laboratory production"), Lund (1976) and B. R. Thomas, Biochimie, 55 (1973) 1325). (From Ref.[1])

ing clear to an increasing extent that for ionic surfactants, electrostatic interactions to a high degree control micelle formation and stability as well as the stability ranges of different liquid crystalline phases. An account of the electrostatics of amphiphilic aggregates is given in Sect. 6 where it is shown that many features can be rationalized on the basis of an unexpectedly simple model.

We have recently presented a rather comprehensive account[1] of micelle formation. The present article is more chemically oriented and contains a complementary rather than repetitive description of the subject. For example, for information on physico-chemical methods amenable for the study of micelle formation the reader is referred to the rather extensive enumeration in Ref.[1].

Section 2 is a broad and schematic review of a voluminous literature and it is evident that a full citation of the original literature is out of question. Several aspects considered have been well described in books by Shinoda et al.[2], by Mukerjee and Mysels[3] and by Tanford[4] as well as in other reviews[5-18]. For reference to published work, as well as to convenient tabulatory presentations, the reader is referred to these presentations; many adequate references are also listed in Ref.[1].

The existing view of a micelle has evolved during a long period of time and it is based on a wide range of experimental approaches. The basic ideas of the roughly spherical micelle stem from the pioneering work of Hartley[19, 20]. In recent years, especially various spectroscopic techniques have considerably enriched our insight into the inner structure of amphiphile aggregates. An attempt to illustrate the structure of a typical micelle is made in Fig. 1.2.

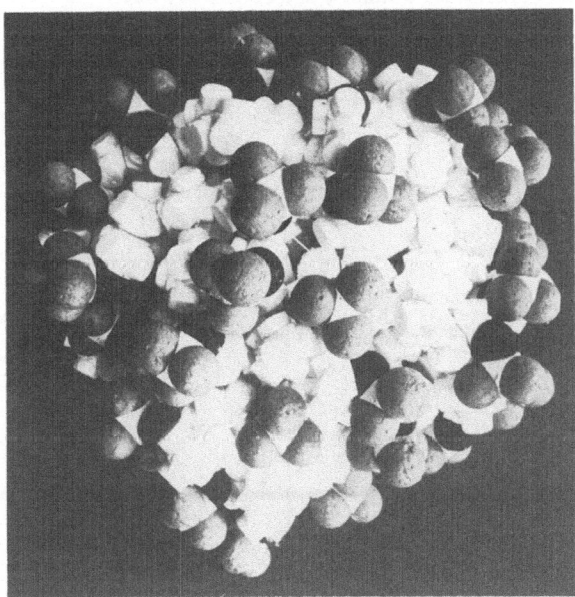

Fig. 1.2. Space filling model of a micelle of SDS. The model was constructed by Hans Gustavsson using the same system as the one in Fig. 1.1. (From Ref. [1])

2 Phenomenological Description of Amphiphile Solutions

2.1 Concept of Critical Micelle Concentration CMC

It was observed some fifty years ago that physico-chemical properties of surfactant solutions may show a peculiar concentration dependence[21, 22]. From such observations, which by now apply to a large number of macroscopic and microscopic quantities (examples are given in Fig. 2.1) the concept of a *critical micelle concentration* CMC has evolved. At low concentrations many physico-chemical properties such as self-diffusion coefficients, activities, turbidity, conductance, surface tension and NMR spectral features indicate that there is no appreciable aggregation of the surfactant. Above the CMC, these properties change in a way indicating that an extensive association to large aggregates is commencing. The CMC concept has an exact meaning within the so-called *phase separation model* of micelle formation. This model considers micelle formation as analogous to a phase separation and the CMC is then the saturation concentration of the amphiphile in the monomeric state and the micelles the separated pseudophase. It is now generally recognized that the phase separation model only provides an approximate description and that the CMC thus has no strict meaning. However, the CMC concept is so useful in giving a characterization of the self-association pattern that it will certainly continue to be extensively used[3].

The CMC is usually determined experimentally by plotting some property as a function of concentration and extrapolating the results at low and high concentrations to an intersection point. It is evident that the value obtained will depend on the type of representation used as well as on the physico-chemical quantity considered. For surfactants with low CMCs, micelle formation begins abruptly and the uncertainties involved are rather small while this is not always the case for surfactants with high CMCs. When the variation of CMC with chemical or physical factors is considered it is often essential to use CMC values obtained in a consistent way.

A large number of experimental techniques can be utilized to determine CMC values, the most popular ones being investigations of electrical conductance, surface tension and the *solubilization* of a compound having a low solubility in water. The

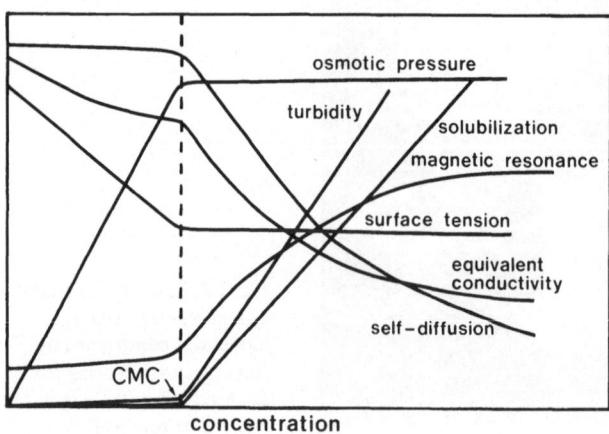

Fig. 2.1. Schematic representation of the concentration dependence of some physical properties for solutions of a micelle-forming amphiphile

surface tension method is valuable in the presence of added electrolyte as well as for compounds with low CMCs but may be quite sensitive to impurities even in trace amounts. The solubilization method may be subject to considerable error because the added compound may participate in the aggregation and thus influence the CMC. Various NMR techniques have recently been introduced for the study of micelle formation and these are useful in that they may give an unambiguous picture of the micelle formation on a molecular level.

A most valuable compilation of CMC data has been presented by Mukerjee and Mysels[3] who in addition to an extensive tabulation of CMCs for various systems discuss the CMC concept as well as plot procedures and experimental techniques for CMC determination. In Table 2.1 we give selected examples of CMC values taken from Ref.[3].

Mukerjee[6, 23, 24] has in detail considered the extraction of CMC values from experimental data and suggested preferred representations of data. In the case of conductance it is preferrable to plot the differential conductance against concentration while in the case of NMR chemical shifts, the product of chemical shift and concentration should be plotted versus concentration. Mukerjee also discusses how micelle formation, i.e., the cooperative aggregation into large aggregates, can be distinguished from other types of step-wise aggregation. As he demonstrates, several

Table 2.1. Selected critical micelle concentrations at 25 °C (mainly from Ref.[3])

Amphiphile	CMC[a]
Sodium tetradecylsulfate	$2.1 \cdot 10^{-3}$ M
Sodium dodecylsulfate	$8.3 \cdot 10^{-3}$ M
Sodium decylsulfate	$3.3 \cdot 10^{-2}$ M
Sodium octylsulfate	$1.33 \cdot 10^{-1}$ M
Dimethyldodecylamineoxide	$2.1 \cdot 10^{-3}$ M
Dodecylammonium chloride	$1.47 \cdot 10^{-2}$ M
Dodecyltrimethylammonium chloride	$2.03 \cdot 10^{-2}$ M
Sodium octanoate	$4 \cdot 10^{-1}$ m
Sodium nonanoate	$2.1 \cdot 10^{-1}$ m
Sodium decanoate	$1.09 \cdot 10^{-1}$ m
Sodium undecanoate	$5.6 \cdot 10^{-2}$ m
Sodium dodecanoate	$2.78 \cdot 10^{-2}$ m
Sodium p-octylbenzenesulfonate	$1.47 \cdot 10^{-2}$ m
Sodium p-dodecylbenzenesulfonate	$1.20 \cdot 10^{-3}$ m
Decyltrimethylammonium bromide	$6.5 \cdot 10^{-2}$ m
Dodecyltrimethylammonium bromide	$1.56 \cdot 10^{-2}$ m
Hexadecyltrimethylammonium bromide	$9.2 \cdot 10^{-4}$ m
$CH_3(CH_2)_9(OCH_2CH_2)_6OH$	$9 \cdot 10^{-4}$ M
$CH_3(CH_2)_9(OCH_2CH_2)_9OH$	$1.3 \cdot 10^{-3}$ M
$CH_3(CH_2)_{11}(OCH_2CH_2)_6OH$	$8.7 \cdot 10^{-5}$ M
$CH_3(CH_2)_7C_6H_4(CH_2CH_2O)_6OH$	$2.05 \cdot 10^{-4}$ M
Dodecylpyridinium chloride	$1.47 \cdot 10^{-2}$ M
Potassium perfluorooctanoate	$2.88 \cdot 10^{-2}$ m

a   In moles/dm^3 (M) or moles/kg H_2O (m)

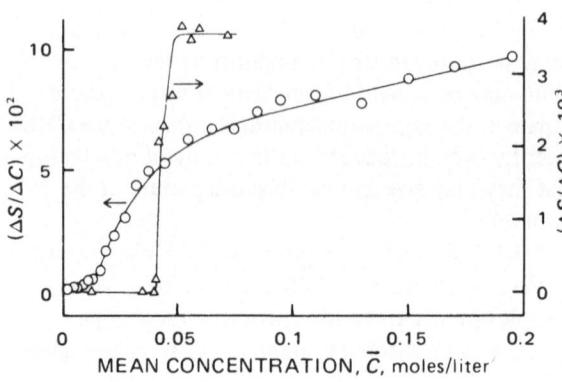

Fig. 2.2. The solubility of naphthalene in aqueous solutions of sodium cholate (○) and of Orange OT in solution of sodium decanesulfonate (△). The ratio between the increment in solubility (ΔS) and in concentration (ΔC) is plotted. The sodium decanesulfonate solution displays the behavior typical of micelle formation with a well-defined CMC. (From Ref.[24])

examples may be found where CMC values have been assigned to noncooperative association as well as to dimerization. An example of plots for micelle forming and nonmicelle forming amphiphiles is shown in Fig. 2.2.

2.2 Variation of CMC with Chemical Structure

Hydrocarbon Chain Length

For single alkyl chain surfactants, which are of principal interest in this treatment, the factor primarily determining the CMC is the size of the hydrophobic part. The dependence of the CMC on the number of carbon atoms (n_C) in the alkyl chain can for many classes of amphiphiles be written[2]

$$\log \text{CMC} = a - bn_C \tag{2.1}$$

where a and b are constants. Typical plots are given in Fig. 2.3. As exemplified there, b is about 0.5 (using log base 10) for many nonionic and zwitterionic systems while it is lower (generally $0.29 - 0.30$) for ionic surfactants. In the presence of added salt, b is increased for the ionic amphiphiles.

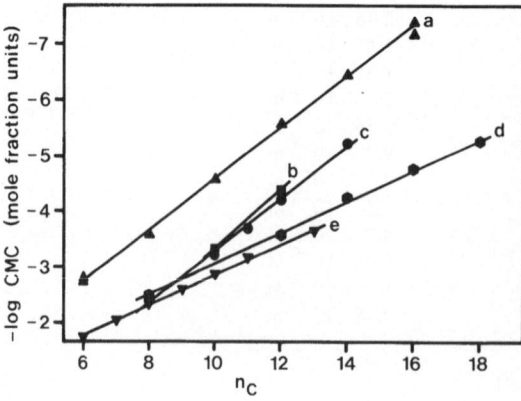

Fig. 2.3. Plots of log CMC (in mole fraction units) versus n_C, the number of carbon atoms in the alkyl chain. (Temperature in general 25 °C). *a.* Alkyl hexaoxyethylene glycol monoethers (Data from Ref.[113]). *b.* Alkyl trimethylammonium bromides in 0.5 M NaBr (Data from Refs.[28, 114]). *c.* N-alkyl betaines (Data from Ref.[115]). *d.* Sodium alkyl sulfates (40 °C) (Data from Ref.[116]). *e.* Sodium alkylcarboxylates (Data from Ref.[41])

Hydrocarbon Chain Structure

Chain branching, as well as introduction of double bonds, has been observed to lead to an increased CMC in comparison with the corresponding n-alkyl compounds. For sodium alkyl sulfates, the CMC increases as the sulfate group is moved from the terminal position. For alkali alkanoates, for example, the CMC has been found to increase by a factor of 3–4 on introduction of one double bond. Addition of a benzene ring leads to a CMC lowering corresponding to 3.5 methylene groups as inferred for example from a comparison between sodium alkylsulfonates and sodium alkylbenzenesulfonates. Substitution of polar groups in the alkyl chain is accompanied by an increased CMC. The introduction of a −OH group has been observed to increase the CMC by a factor of 4 corresponding to the removal ot two methylene groups.

Fluorocarbon Chains

Partial or total fluorination has striking effects on the CMC. Thus while perfluorinated compounds have much lower CMCs than the corresponding hydrocarbon surfactants (as an example the CMC is 0.030 M for sodium perfluorooctanoate and 0.38 M for sodium octanoate), partial fluorination increases the CMC. The strong deviation from an ideal behavior is illustrated in Fig. 2.4[25]. A markedly nonideal behavior is also found in mixtures of hydrocarbon and fluorocarbon surfactants[26].

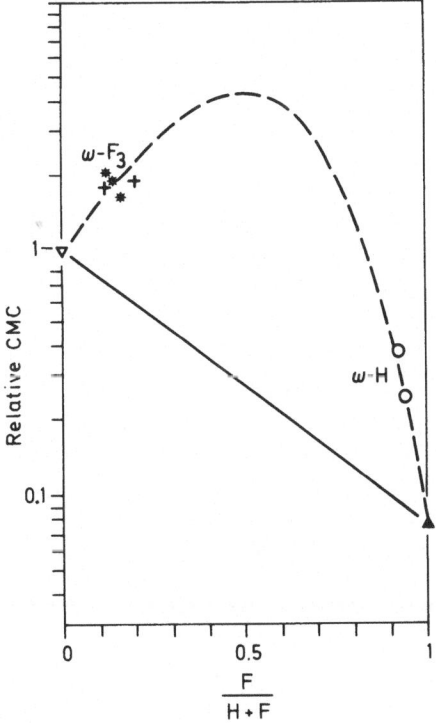

Fig. 2.4. CMC of various hydrocarbon, fluorocarbon and partially fluorinated surfactants as a function of the fluorine-to-hydrogen ratio. CMCs are given relative to those of the hydrocarbon surfactants. Solid line is a suggested ideal behavior (From Ref.[25])

Polar Head-Group

In regards to the nature of the polar head-group, the main influence comes from its charge so that CMCs are, for a given alkyl chain length, much lower for nonionic than for ionic surfactants. Zwitterionic surfactants generally fall in between of these groups. For ionic surfactants, there are rather small differences between different head groups. For n-dodecyl surfactants the CMCs are (counterion in parentheses)

8 mM for $-SO_4^-$ (Na^+),

15 mM for NH_3^+ (Cl^-),

20 mM for N^+ $(CH_3)_3$ (Cl^-),

10 mM for $-SO_3^-$ (Na^+) and

12 mM for $-CO_2^-$ (K^+).

For nonionics, the CMCs may be markedly affected by the size and nature of the hydrophilic group. For ionic surfactants, the CMC is increased on introduction of additional ionic groups; for example, the CMC of $C_{10}CH$ $(COOK)_2$ is 0.13 M while it is 0.024 M for $C_{11}COOK$.[1]

Counterion

It is mainly the valency of the counterion which influences the CMC while other factors have small effect for simple inorganic ions. The CMCs of several dodecylsulfates with divalent counterions $(Ca^{2+}, Mg^{2+}, Pb^{2+}, Zn^{2+}$ etc.) are around 2 mM while the alkali dodecylsulfates have CMCs of ca. 8 mM. Among the alkali ions, differences in CMC are small and it has shown to be no easy matter to establish them experimentally. Mukerjee et al.[27] noted that the CMC of alkali dodecylsulfates increases slightly with decreasing counterion atomic number while with tetraalkylammonium ions as counterions, the CMC decreases with increasing ion size. For cationics, the counterion effects are more pronounced. For n-dodecyltrimethylammonium salts the CMC follows the sequence[28] $NO_3^- < Br^- < Cl^-$ and correspondingly for hexadecyltrimethylammonium salts the CMC is lower[3] with Br^- than with Cl^-. For decylammonium salts, the CMC sequence is $NO_3^- < Br^- < Cl^-$ and for alkylpyridinium halides the CMC decreases appreciably with increasing counterion size[3]. Organic anions have in a number of cases been found to decrease the CMC markedly. For hexadecyltrimethylammonium salts, the CMC is lowered by a factor of 5–10 on substitution of salicylate or certain other substituted benzoates for bromide[29] and low CMCs have also been found for tetradecylpyridinium salts with organic counterions (e.g., alkylbenzenesulfonates and alkylsulfonates)[30].

---

1 C_n denotes an n-alkyl chain with n carbon atoms.

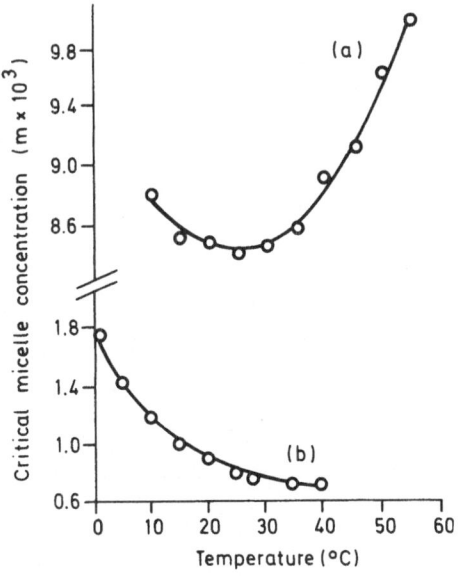

Fig. 2.5. Variation of CMC with temperature for (a) $CH_3(CH_2)_{11}SO_4Na$ and (b) $CH_3(CH_2)_9(OCH_2CH_2)_5OH$. (From Ref. [13])

2.3 Variation of CMC with Intensive Parameters

Temperature

A general feature of the CMC is the weak temperature variation compared to most other chemical association phenomena[31]. As regards the details, several types of behavior have been observed: The CMC may increase or decrease with increasing temperature or it may show a pronounced minimum. Examples of temperature dependences are shown in Fig. 2.5.

Pressure

The pressure dependence of the CMC is weak even up to quite high pressures as exemplified in Fig. 2.6 for sodium dodecanoate[32].

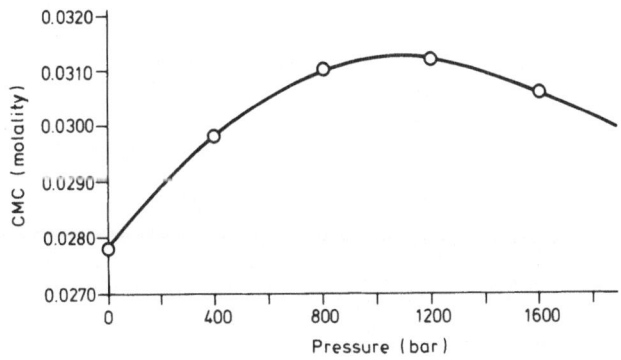

Fig. 2.6. Variation of CMC with pressure for sodium dodecanoate (From Ref. [32])

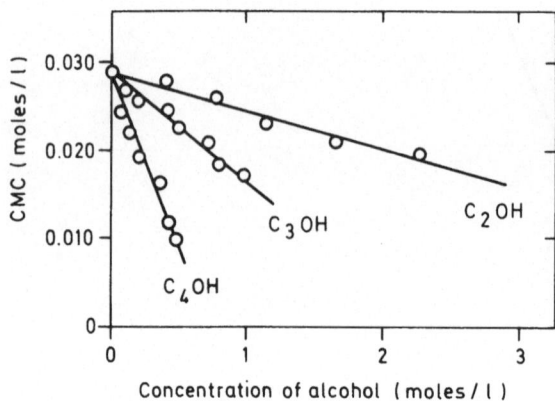

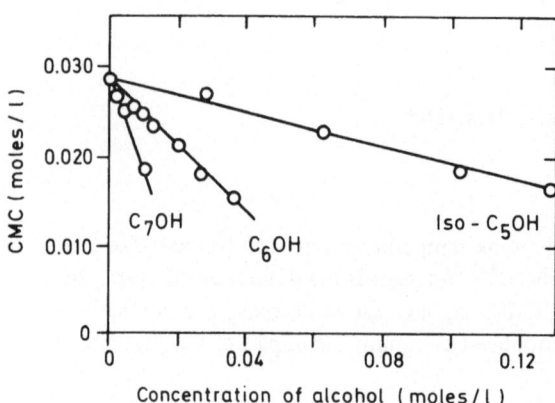

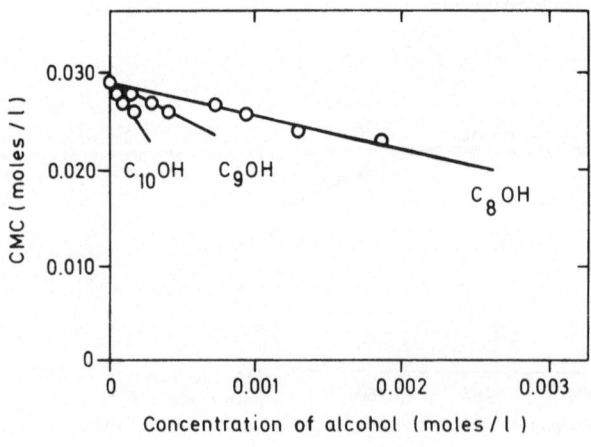

Fig. 2.7. The effect of alcohols on the CMC of potassium dodecanoate at 10 °C. (From Ref.[2])

Added Simple Electrolyte

While for nonionic systems, the effect of inorganic salt on CMC is small, it is large
for ionic surfactants and this difference is expected since the micelle – small ion interaction must be completely different in the two cases. The decrease in CMC for ionic
systems corresponds usually to a linear relation between the logarithm of the CMC
and the total counterion concentration, if a salt of the counterion is added. For divalent ions, the slope of the plot of log CMC versus counterion concentration is half
of that of monovalent ions.

Added Nonelectrolytes and Amphiphiles

As expected, the influence of added nonelectrolytes can be quite different depending
on whether the added compound is likely to be located in the micelles or in the intermicellar solution. The effect of normal alcohols has been studied in detail for
potassium dodecanoate; the CMC is lowered for all alcohols studied but the effect
increases considerably in going from ethanol to decanol (cf. Fig. 2.7). Hydrocarbons,
like cyclohexane, n-heptane, toluene, and benzene, have been found to lower the
CMC for many surfactants. Strongly hydrophilic substances, like dioxane and urea,
have small and complex effects. At higher concentrations they markedly increase
the CMC or even inhibit micelle formation. Addition of another similar surface-
active agent generally gives a CMC in between the CMCs of the two surfactants.

2.4 Solubility of Amphiphiles. The Krafft Phenomenon

It is an everyday experience that calcium soaps have a low aqueous solubility while
many surfactants with monovalent counterions have an extremely high solubility.
As for other compounds, the solubility is to a great extent given by conditions in
the solid phase, but for surfactants the strongly cooperative association produces
the peculiar temperature dependence of solubility schematized in Fig. 2.8. The so-

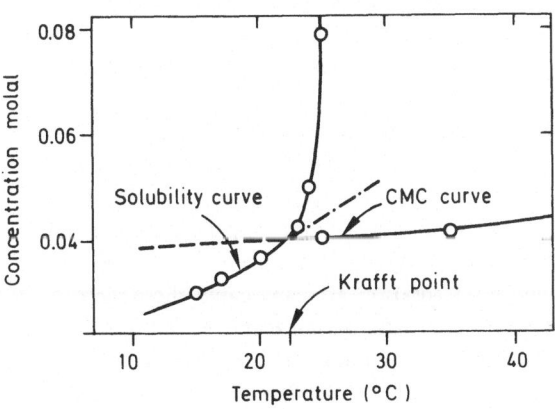

Fig. 2.8. Temperature dependence of surfactant solubility
in the region of the Krafft
point. (From Ref.[2])

13

lubility is low at low temperatures but increases dramatically around a certain temperature referred to as the *Krafft point*. The explanation is that at low temperatures the low monomer solubility determines the total solubility, while at higher temperatures when the monomer solubility has reached the CMC it is determined by the micelle solubility which is much higher. The Krafft phenomenon is often characterized by a *critical micelle temperature*, CMT[33]. The Krafft point shows a considerable variation with both alkyl chain length and counterion and may be strongly influenced by even small amounts of impurities. Examples of Krafft points are 20 °C for sodium dodecylsulfate, 27 °C for hexadecyltrimethylammonium bromide, 36 °C for sodium dodecanoate, 53 °C for sodium tetradecanoate and 8 °C for potassium tetradecanoate. The variations with the alkyl chain length may be qualitatively understood from differences in monomer solubility while the large variations between alkali counterions must involve ionic interactions in the solid state. That the conditions in the solid state are decisive is also shown by the close correlation between the Krafft temperature and the melting of the hydrocarbon chain as evidenced by the appearance of liquid crystalline phases. From phase diagrams of two-component systems of surfactant and water it can be inferred that the saturated isotropic solutions slightly above the Krafft point often are in equilibrium with a hexagonal mesophase instead of the solid hydrated surfactant.

2.5 Physico-Chemical Properties of Surfactant Solutions. Thermodynamic and Transport Properties

The formation of aggregates of colloidal dimensions in amphiphile solutions was clearly established in early light scattering studies. Due to the micelles, there are large fluctuations in the refractive index in the solution resulting in a high turbidity. By measuring the turbidity of a series of solutions at different concentrations, light scattering has been used extensively to obtain information on micelle sizes[9]. A most valuable complement to classic light scattering techniques has in recent years been *quasi-elastic light scattering* spectroscopy (QLS)[33]. In this technique, the temporal fluctuations in the intensity of the scattered light are used to obtain the translational diffusion coefficients of the micelles. The fluctuations arise from the brownian movement of the micelles and can be monitored by measuring the autocorrelation function of the intensity of the scattered light. Measurements of the dissymmetry of the scattered light can be very useful for detecting and identifying changes in micelle shape[34].

The *viscosity* starts to increase above the CMC and it is well established that the viscosity of a colloidal solution can give information on size and shape of the particles. From studies of the viscosity as a function of micellar concentration, the intrinsic viscosity may be obtained by extrapolation. The intrinsic viscosity depends on a shape factor, and the micelle specific volume and viscosity studies are therefore used to determine micelle shape and hydration. In many cases, these factors appear to be quite constant over a wide concentration range above the CMC. In other cases, such as hexadecyltrimethylammonium bromide (Fig. 2.9), dramatic increases in viscosity are observed at higher concentrations[35]. Studies of surfactants with low

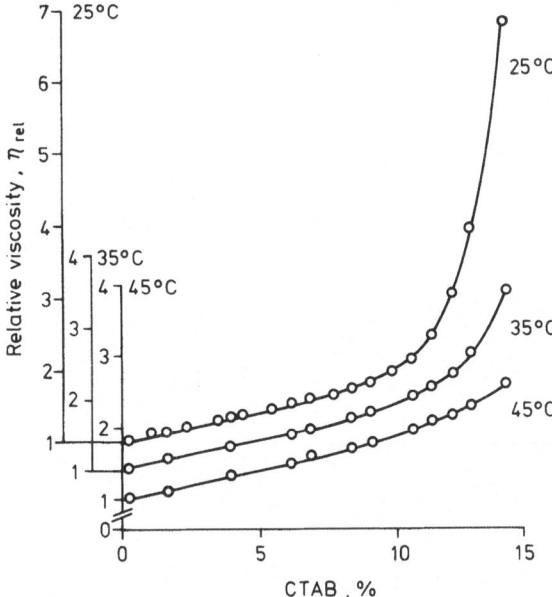

Fig. 2.9. Relative viscosity as a function of CTAB concentration (in weight percent) at 25 °C, 35 °C and 45 °C. (From Ref.[35])

CMCs in the absence of added electrolyte may be greatly influenced by electroviscous effects; marked decreases in intrinsic viscosity on electrolyte addition have been observed in many cases[36]. Peculiar and highly interesting rheological properties of surfactant solutions include observations of strongly non-Newtonian behavior as well as of viscoelasticity; these are yet incompletely understood.

The *electrical conductance* shows a weaker concentration dependence above than below the CMC corresponding to a decrease in the equivalent conductance (Fig. 2.10). The transport number of the surfactant ion rises sharply at the CMC while that of the counterion may become negative. This as well as *electrophoretic mobilities* may yield information on micellar charge. At high concentrations, *conductance anisotropies* have been observed for flowing systems. This, as well as *flow birefringence*, is useful for the demonstration of nonspherical micelle shape.

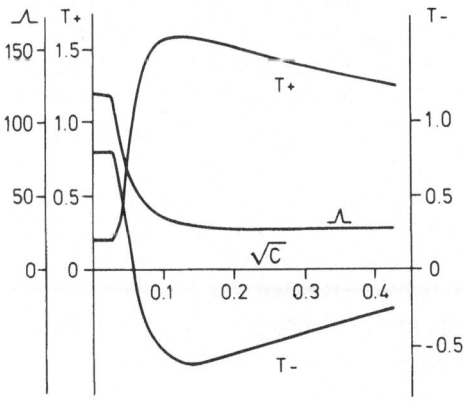

Fig. 2.10. Equivalent conductance (Λ) and transport number of anion (T−) and cation (T+) for CTAB solutions at 35 °C plotted versus the square root of surfactant molarity. (From Ref.[13])

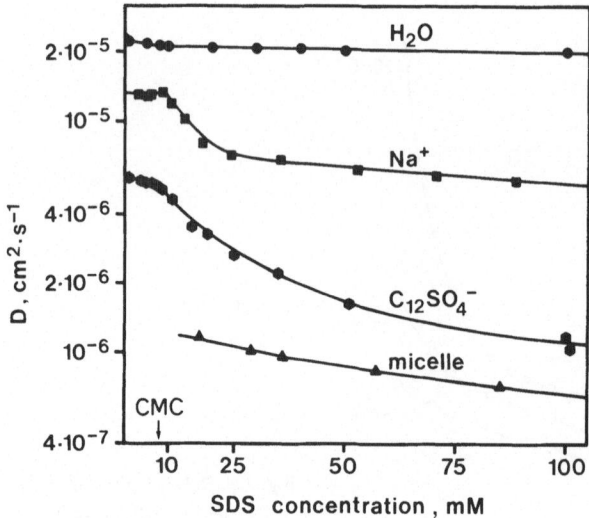

Fig. 2.11. Self-diffusion coefficients of water (●), sodium ions (■), dodecylsulfate ion (●) and micelles (▲) in SDS solutions. Data from Refs.[37, 110–112]

The *self-diffusion* of the individual components is strongly affected by the formation of micelles in the solution. This applies to the surfactant, the counterion, the water, and to solubilized molecules. As illustrated in Fig. 2.11 for sodium dodecyl sulfate, surfactant and counterion diffusion are very weakly dependent on concentration below the CMC while a marked decrease in the micellar region is observed for the surfactant and a less marked one for the counterion[37]. Water diffusion shows a stronger concentration dependence below the CMC than above it. Self-diffusion studies using radioactive tracers have been performed to obtain information on CMC, on counterion binding, on hydration and on intermicellar interactions and shape changes.

For the testing of theoretical models, it is most important to have access to relevant thermodynamic data. ΔG^{θ} values are obtained from data on micellization equilibria; for long-chain surfactants, the CMC gives ΔG^{θ} with good precision. Direct determinations of ΔH^{θ} using *calorimetric* methods have given micellization enthalpies; these are quite small and may be positive or negative. They lie in the range -5 to $+5$ kJ $\cdot$ mol^{-1} in most cases. Combination of ΔG^{θ} and ΔH^{θ} data has shown that micellization is accompanied by a distinct entropy gain. Heat capacity data pertaining to micelle formation are not abundant but in the cases investigated ΔC_p^{θ} has been found to be markedly negative. Partial molal excess quantities of the binary system sodium octanoate-water have been determined by Danielsson et al.[38] and as shown in Fig. 2.12 both the partial excess enthalpy and the partial excess entropy change considerably in a wide concentration range below the CMC.

From studies of the concentration dependence of *density* and *isentropic compressibility coefficients,* the apparent molar volume and the isentropic apparent molar compressibility may be obtained above and below the CMC. Such studies have recently been performed for several systems by Brun, Høiland and Vikingstad[32, 39–41] who deduced the change in partial molar volume and compressibility on micelle formation. This gives information on the counterion hydration and the packing of the hydrocarbon chains in the micelles.

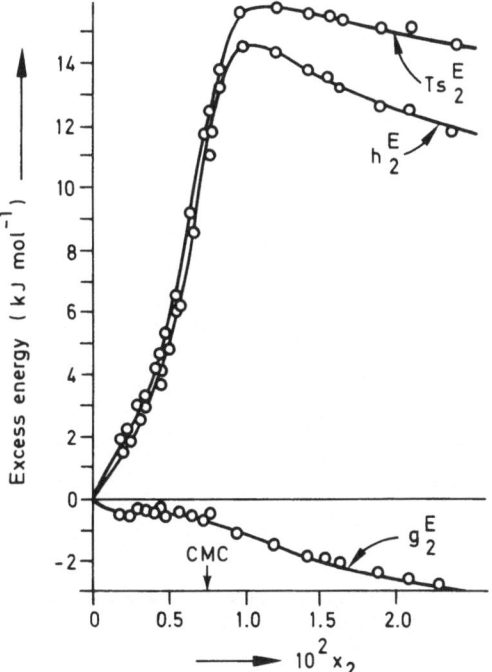

Fig. 2.12. The partial excess free energy, g_2^E, the partial excess enthalpy, h_2^E, and the partial excess entropy times temperature, Ts_2^E, for aqueous solutions of sodium octanoate. (From Ref.[38])

The *counterion activity* of surfactant solutions rises linearly with concentration up to the region of the CMC while it takes on a slower increase above the CMC[42]. As an example we show in Fig. 2.13 measurements by Brun et al.[43] using a cation exchange electrode. Larsen et al.[44, 45] used bromide and chloride specific electrodes for solubilization studies; in 0.1 M hexadecyltrimethylammonium chloride or bromide solutions, 76—77% of the counterions are bound, a figure changing only slightly

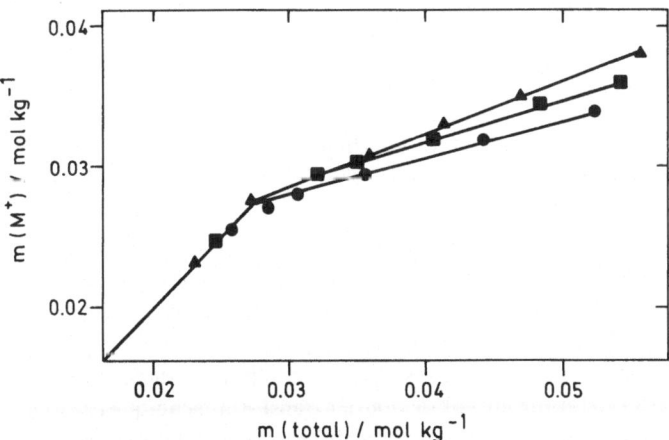

Fig. 2.13. Concentration of free counterions for sodium (•), potassium (■) and tetramethyl-ammonium (▲) dodecanoates as a function of dodecanoate concentration. (From Ref.[43])

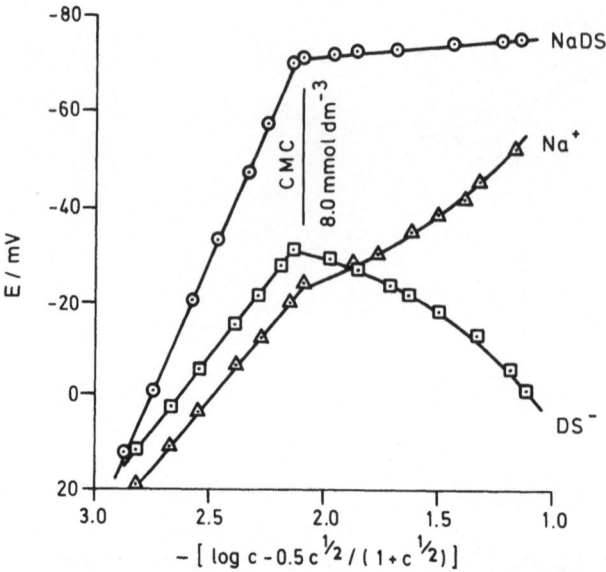

Fig. 2.14. Results from measurements of the activity of sodium dodecylsulfate, of sodium ions and of dodecylsulfate ions in solutions of sodium dodecylsulfate. (From Ref.[46])

on solubilization. Of great significance is the development of surfactant specific electrodes which has been achieved recently by Cutler et al.[46]. As can be seen in Fig. 2.14 for the case of dodecylsulfate, *amphiphile activity* changes from an increase with concentration to a decrease at the CMC. The water vapour pressure of sodium octanoate solutions is reduced at a lower rate above than below the CMC[38] and for nonionics it has been observed that a micelle forming amphiphile gives a smaller reduction of water activity than a compound not forming micelles[47]. The activity of a solubilizate (pentanol in sodium octanoate solutions, for example) is considerably decreased on micelle formation[38]. The osmotic pressure of surfactant solutions is approximately proportional to concentration below the CMC, but changes very slightly above this concentration[48].

2.6 Physico-Chemical Properties of Surfactant Solutions. Spectroscopy

The formation of micelles will influence the local molecular interactions for the components. This in turn shows up as a change in a number of spectroscopic parameters. Thus it is today possible to probe several different aspects of the molecular organization in micellar systems by spectroscopic methods.

The most generally applicable tool for the study of amphiphile systems is *nuclear magnetic resonance* (NMR). With the advent of Fourier transform techniques as well as high field superconducting magnets, NMR studies are feasible at submillimolar concentrations for many nuclei and thus highly appropriate for investigations of micelle formation. Studies of the hydrophobic part of the amphiphile may be made

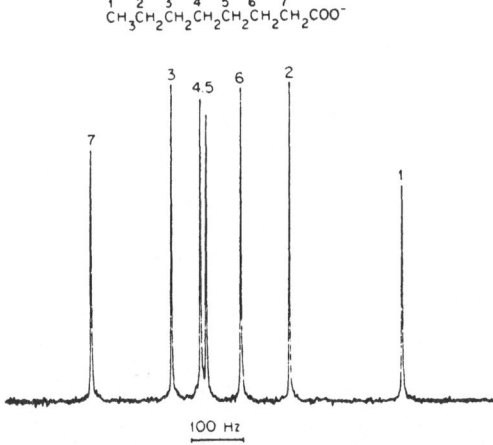

$$\overset{1}{C}H_3\overset{2}{C}H_2\overset{3}{C}H_2\overset{4}{C}H_2\overset{5}{C}H_2\overset{6}{C}H_2\overset{7}{C}H_2COO^-$$

Fig. 2.15. ^{13}C NMR spectrum of 1 M sodium octanoate. (From Ref.[49])

by means of ^{1}H, ^{2}H, ^{13}C and ^{19}F NMR. The chemical shifts of ^{13}C and ^{19}F NMR are sizeable and it is straightforward to resolve and assign a large number of different signals. This is exemplified by the ^{13}C spectrum of octanoate in Fig. 2.15. As in most NMR studies the spectral parameters are population weighted averages of the different environments the ion or molecule samples. The ^{13}C chemical shift is relatively independent of surfactant concentration below the CMC, but begins to change in the region of the CMC[49–51] (Fig. 2.16). The magnitude of the ^{13}C chemical shift changes in an alkyl chain are larger in the middle of the chain than at either end. ^{19}F chemical shifts[52] follow the same type of concentration dependence as depicted for ^{13}C.

The relaxation of ^{19}F is often appreciably different in H_2O and D_2O for the monomeric surfactant while this difference is almost completely eliminated in the micelles[53, 54]. ^{1}H NMR has a very good sensitivity but a low resolution due to small chemical shifts, which reduce its applicability. For surfactants with aromatic groups appreciable shift effects are observed[55] and on solubilization of aromatic compounds in surfactant micelles one observes much larger shifts for groups close to the polar group than for remote groups[56]. Several simple counterions are well suited for study by NMR and it has been observed in several cases that counterion quadrupole rela-

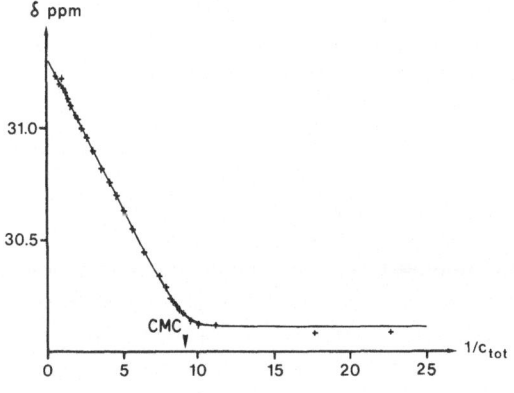

Fig. 2.16. ^{13}C NMR chemical shift of carbon 4 (numbering starting from the polar head) in nonylammonium bromide as a function of the inverse amphiphile concentration. A positive shift is downfield. Solid line calculated with an aggregation number of 33. (After Ref.[51])

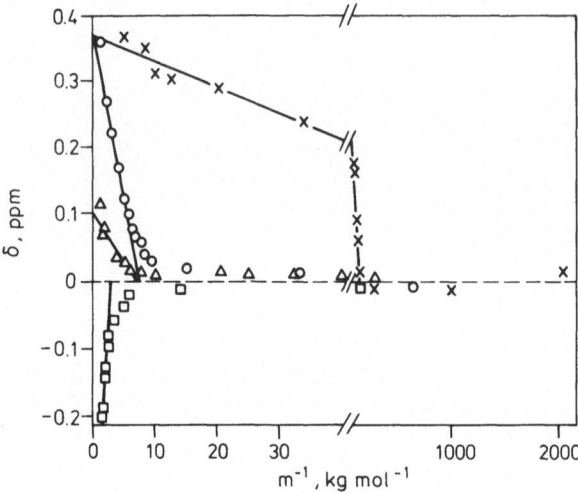

Fig. 2.17. Plots of ^{23}Na$^+$ NMR chemical shifts versus the inverse concentration of sodium octanoate (□), sodium octylsulfate (○), sodium octylsufonate (△) and sodium dodecylsulfate (X). A positive shift is upfield. (From Ref.[57])

xation rates and chemical shifts are rather insensitive to surfactant concentration when this is low but strongly concentration dependent above the CMC[57]. As can be inferred from Fig. 2.17 the changeover is very distinct for long-chain surfactants but much less so for short-chain ones. At a change in micellar shape from spherical to cylindrical the counterion quadrupole relaxation rate has been observed to increase[58]. Solubilization in certain cases leaves the counterion quadrupole relaxation rate unaffected while in others it is decreased considerably. For certain ions like ^{133}Cs$^+$ there is a rather large chemical shift difference between H$_2$O and D$_2$O solutions. For cesium octanoate solutions, the shift difference has the same value as for infinite dilution up to very high surfactant concentrations[59].

Studies of the water ^{1}H relaxation as a function of surfactant concentration have given a slower rate of change above than below the CMC[60, 61]. For solutions containing paramagnetic counterions large changes in ^{1}H water relaxation occur on counterion binding[62].

Micellar solutions are in most cases isotropic from an NMR point of view, i.e., due to rapid molecular motion the various NMR effects are averaged to their isotropic mean values. This motional averaging eliminates from the spectrum a lot of useful information which one should have for the corresponding anisotropic system. NMR spectra of anisotropic surfactant-water mesophases therefore contain extensive complementary information which in view of great structural similarities can be applied, although with due caution, also to micellar solutions. Quadrupole splittings of both counterions[63] and water[64, 65] and alkyl chain[66] deuterons, therefore, are prominent features of NMR spectra of mesophases. In special cases it has been possible to obtain quadrupole-split spectra even for optically isotropic micellar solutions, e.g. ^{2}H spectra of deuterated benzene and cyclohexane in concentrated solutions of hexadecyltrimethylammonium bromide[67].

Electron spin resonance (ESR) spectroscopy is much more sensitive than NMR but one is, of course, strongly hampered by the requirement of unpaired electrons. The study of nitroxide-labelled compounds, both surfactants and solubilizates, has become a standard method in surfactant science[68]. The magnitude of the hyperfine splitting of a compound in a micelle can vary between values typical of nonpolar environments to values found for aqueous solutions depending on the location of the nitroxide group in relation to polar and nonpolar groups. Studies of paramagnetic counterions are also often informative. The correlation time of the VO^{2+} ion was found to change only slightly on the ion's binding to dodecylsulfate micelles[69].

In *Raman spectra*, the C–C stretching frequency is sensitive to the conformation of the alkyl chain. With increasing surfactant concentration an increase of the trans bands was observed at the expense of that of the gauche bands, but somewhat unexpectedly the change occurs well below the CMC while micelle formation seems not to be associated with appreciable spectral effects[70−72]. Changes in C–H stretching vibrations, which for simple systems correlate with solution polarity, are observed to occur over a wide concentration region up to concentrations considerably above the CMC[70, 72].

For many simple surfactant systems, the methods of *electron and fluorescence spectroscopy* are not directly applicable, but in quite a few cases either the surfactant or a solubilized molecule displays useful light absorbing and/or fluorescing properties. However, it is more frequently so that measurements are made on a spectroscopic probe added in small amounts to the system of interest.

For alkylpyridinium iodide micellar solutions, there is a strong charge transfer band with the bound iodide counterion as donor and the pyridine ring as acceptor. By comparing with corresponding spectra of ion pairs in different solvents one obtains information on the local environment at the micellar surface[73]. (The polarity was expressed in terms of a so-called effective dielectric constant). Amphiphiles with a benzene ring also show an UV absorption[74−76]. UV spectra of solubilized species like benzene, naphthalene and pyrene have been extensively studied and compared with spectra of the compound in reference solvents to provide an estimate of the polarity in the vicinity of the solubilizate.

The fluorescence of probe molecules often shows distinct changes on solubilization in micelles. A characteristic intensity ratio in the vibronic fine structure of pyrene monomer fluorescence is, for a large number of surfactant micelles, closer to that for aqueous than for hydrocarbon solutions[77]; it is also markedly dependent on surfactant end-group. The pyrene monomer fluorescence lifetime is observed to increase sharply from 180 ns in aqueous solution to 350 ns on solubilization in SDS micelles[77]; the fluorescence lifetime is considerably shorter for pyrene in CTAB micelles than in micelles of SDS or $C_{16}C(NH_3)_3Cl$ and it is also strongly affected by added ions[78−81]. The transition from spherical to rod-like micelles is accompanied by an increase in the excimer yield[82].

2.7 Surfactant Aggregation at High Concentrations. Phase Diagrams of Two-Component Systems

The onset of formation of approximately spherical micelles from the monomers occurring within a narrow concentration region is a common phenomenon for a large number of amphiphiles, but as one goes to higher concentrations the physico-chemical properties change rather differently for different systems. As will be discussed in detail in Sect. 4, structural effects which have to be considered include changes in micelle shape and size, changes in hydration, counterion binding and hydrocarbon chain conformation, and intermicellar interactions; it is natural to think that these effects are not independent of each other, but that in most cases two or more of them are coupled to each other.

Although a wide range of variability exists, it is possible to make a rough division of the behavior into two groups

a) Physico-chemical properties remain approximately constant or vary at a constant ratio (depending on the property considered) up to very high surfactant concentrations. Both molecular and macroscopic properties show a simple continuation of the properties once above the CMC.

b) A number of physico-chemical properties (but not all; see below) undergo rather dramatic changes in a concentration range well below the saturation limit.

Most surfactants seem to belong to the first group although the behavior may in certain cases change rather drastically with moderate changes in temperature. (Solubilizate and electrolyte may also have dramatic effects). For group (a) amphiphiles, it is, for example, observed that the viscosity changes only slightly with concentration as do the micellar self-diffusion or mutual diffusion coefficients. Also the light scattering intensity shows only moderate changes and ^1H and ^{13}C NMR signals remain very narrow even for extremely concentrated solutions. For a large number of spectroscopic and other properties, a regular concentration dependence is shown which corresponds to approximately constant spectral properties of amphiphile and counterion in the micelles and to a constant ratio of counterions and amphiphile ions in the micelles (Fig. 2.16 and 2.17). Properties displaying such a simple behavior include inter alia ^{13}C and ^{19}F NMR chemical shifts and relaxation times, counterion (e.g., $^{23}Na^+$, $^{35}Cl^-$, $^{133}Cs^+$) NMR chemical shifts and relaxation times, counterion mobility (VO^{2+}) from ESR studies, counterion and amphiphile self-diffusion coefficients and counterion activity. Examples of amphiphiles belonging to group (a) are hexadecyltrimethylammonium chloride, SDS and sodium octylbenzenesulfonate.

For group (b) compounds, which are exemplified by CTAB and sodium oleate, certain molecular and macroscopic parameters change dramatically as one goes from the CMC up to high concentrations. There are distinct changes in the small-angle X-ray scattering[83] and in the light scattering[35], the viscosity changes often immensely[35] (Fig. 2.9), there is an appearance of effects pointing to a strong orientation for flowing systems like flow birefringence, electrical conductance anisotropy and linear dichroism, the ^{13}C and ^1H NMR signals become broad and the micelle self-diffusion is considerably slowed down. Several spectroscopic properties point to changes in the behavior of the micellized amphiphile ions or counterions or in the micelle composition. For example, there are changes in counterion quadrupole relaxation, ^{13}C chemical shifts, pyrene excimer yield and fluorescence depolarization. The rheological properties show particularly large changes and in many cases the solutions show a non-Newtonian behavior and for a number of cases pronounced viscoelasticity has been demonstrated. That the classification of a certain surfactant depends very much on temperature is probably best illustrated by the viscosity of solutions of the hexadecyltrimethylammonium halides. Thus at the lowest temperatures accessible both the chloride, the bromide and the iodide belong to group (b) while at high temperatures all three belong to group (a) (cf. Fig. 2.9). The tendency to give an increasing viscosity is highest with I^- as counterion and lowest with Cl^-.

The appearance of group (b) compounds clearly shows that amphiphile association is not limited to the formation of relatively small spherical micelles but that association may progress considerably to the formation of very large aggregates. However, aggregation does not stop at the saturation limit of a surfactant in water. Thus in the composition range between the optically isotropic micellar solutions and the solid crystalline (or liquid crystalline) pure amphiphile there are typically regions with various lyotropic liquid crystalline phases. As an example we show in Fig. 2.18 the phase diagram for the system dodecyltrimethylammonium chloride-water[84]. The structures of the phases have been determined using low-angle X-ray diffraction techniques[85]. The sequence of the appearance of the different phases for a two-component system of water and a typical micelle-forming amphiphile is generally isotropic solution → cubic phase, type I → hexagonal phase → cubic phase, type II → lamellar phase but for most systems all phases do not occur. Also, the occurrence of a certain phase may depend critically on temperature as well as on minor chemical features[8]. For example, at room temperature, there is a separation out of a cubic phase for dodecyltrimethylammonium chloride while hexadecyltrimethylammonium chloride gives a hexagonal phase. In both cases, the micelles seem

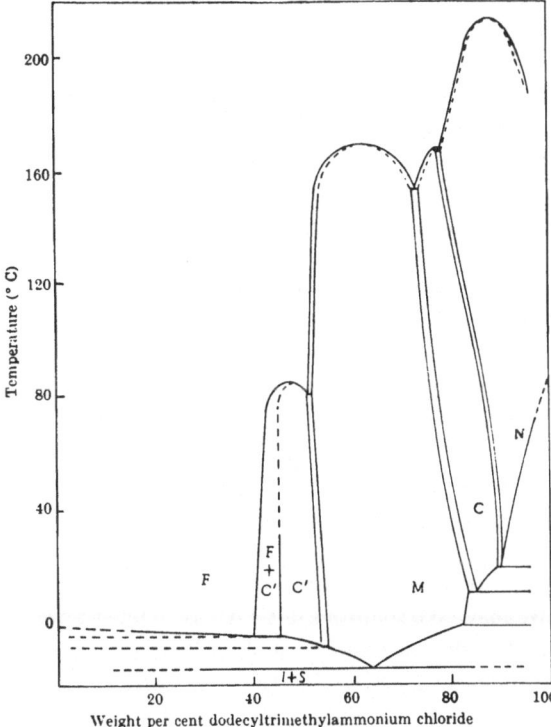

Fig. 2.18. Phase diagram of the dodecyltrimethylammonium chloride-water system. F denotes isotropic solution phase, M normal hexagonal liquid crystal, N lamellar liquid crystal and C' and C cubic liquid crystalline phases. (From Ref.[84])

to retain a closely spherical shape up to high concentrations. CTAB, for which rod-like micelles exist over a wide concentration range, gives a hexagonal phase. According to the phase rule, there should be distinct two-phase areas between the homogeneous phase regions and this is a general observation.

2.8 Solubilization. Phase Diagrams of Three-Component Systems

The solubilization phenomenon, which refers to the dissolution of normally insoluble or only slightly soluble compounds in water caused by the addition of surfactants, is one of the most striking effects encountered for surfactant systems. Solubilization is of considerable physico-chemical interst, such as in discussion of the structure and dynamics of micelles and of the mechanism of enzyme catalysis, and has numerous practical applications, such as in detergency, in pharmaceutical preparations and in micellar catalysis. In biology, solubilization phenomena are most significant, e.g., cholesterol solubilization in phospholipid bilayers and fat solubilization in fat digestion and transport.

The extent of solubilization depends to a large extent both on the solubilizate and surfactant. CTAB micelles may solubilize more than equivalent amounts of several classes of compounds such as benzene, isopropylbenzene, nitrobenzene N,N-dimethylaniline, cyclohexane as well as long-chain alcohols[56]. For sodium octanoate, there is an extensive solubilization of long-chain alcohols while the solubilization of for example p-xylene is moderate and that of octane rather low[86]. The ability to solubilize in general increases strongly with increasing hydrocarbon chain length of the surfactant[2, 87–91]. For a given length of the surfactant, solubilization increases with increasing micelle size[89, 92, 93]. Another interesting observation is that dye (Orange OT) solubilization is considerably reduced on partial fluorination of the surfactant[94]. We have already indicated that solubilization starts at the CMC and in an ideal case the amount of solubilized substance is expected to increase in proportion to the micelle concentration. This applies for many cases but a more complex behavior is frequently encountered as examplified in Fig. 2.19 for decanol solubilization in sodium alkanoate solution[86]. For nonpolar compounds it is probably often appropriate to consider solubilization as a simple partitioning between the micellar nonpolar interior and the aqueous environment in the intermicellar solution. For partly polar solubilizates, this is certainly not a good description since the solubilizate will participate in and influence the surfactant aggregation process. For the case exemplified by decanol solubilization in alkanoate solutions, one might therefore consider solubilization as a formation of mixed micelles. This is evident from the large effect of solubilizates on the CMC mentioned above and is exemplified by the well-known effect of partial hydrolysis of SDS on the concentration dependence of the surface tension.

The influence of solubilization on solution properties may vary from no appreciable effects to very marked ones. We may take solutions of CTAB as an illustrative example of the manifold of possibilities. Here solubilization of cyclohexane has a very small influence on a variety of rheological and spectroscopic properties while addition of aromatic compounds and long-chain alcohols may cause extensive changes.

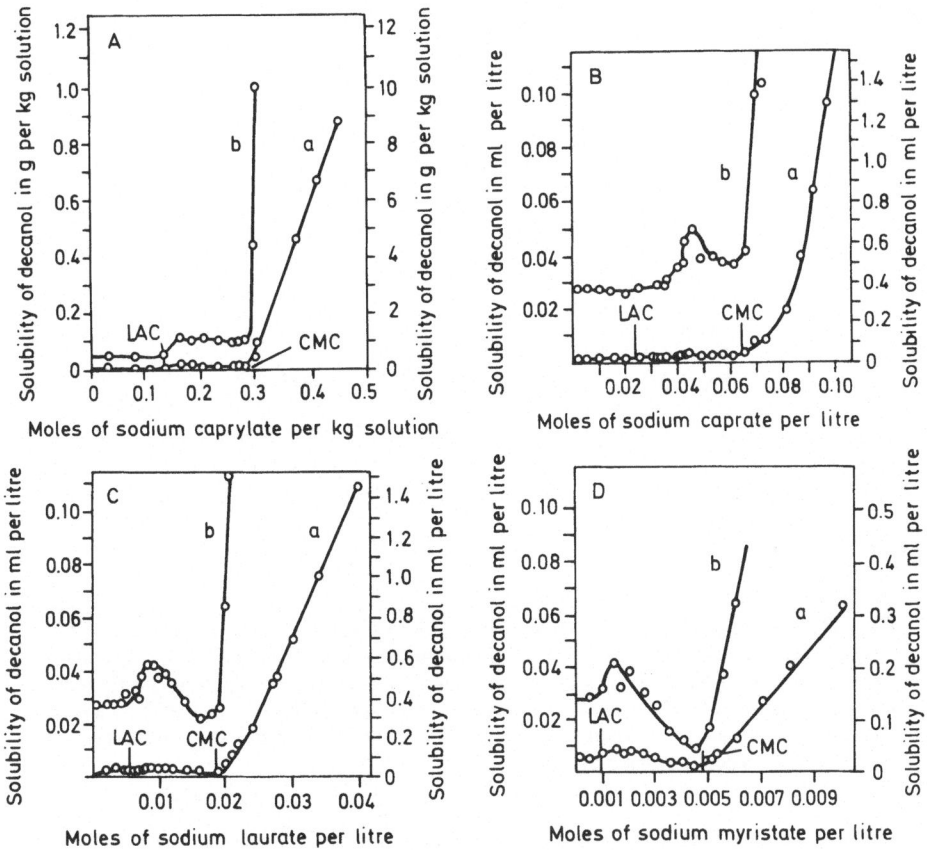

Fig. 2.19 A–D. Solubility of decanol in solutions of (A) sodium octanoate (20 °C), (B) sodium deca-noate (40 °C), (C) sodium dodecanoate (40 °C) and (D) sodium tetradecanoate (40 °C). Curves *a* refer to right hand ordinate axes and curves *b* to left hand ordinate axes. (From Ref.[86])

Rheological properties are particularly sensitive and for some solubilizates, such as α-methylnaphthalene, the solutions may become viscoelastic; the appearence of vis-coelasticity often depends on subtle effects in chemical structure[29]. Certain spec-troscopic features are strongly influenced. Thus the ^{1}H and ^{13}C NMR line widths show large increases, the ^{81}Br$^-$ quadrupole relaxation may be strongly affected and there may be the appearance of linear dichroism, birefringence and conductance anisotropy for flowing systems.

The very complex variation of the amount solubilized, as well as physico-chem-ical properties, with chemical structure of solubilizate and surfactant as well as with surfactant concentration cannot be adequately discussed solely in terms of the ener-getical conditions of the solubilizate in the micelles. Thus one should also consider the conditions in the phase which separates out at the solubilization limit; this is in most cases a liquid crystalline phase. A fundamental basis for a proper understanding of solubilization in surfactant systems is, therefore, a detailed information on phase equilibria in three-component systems surfactant-solubilizate-water. Due in particular

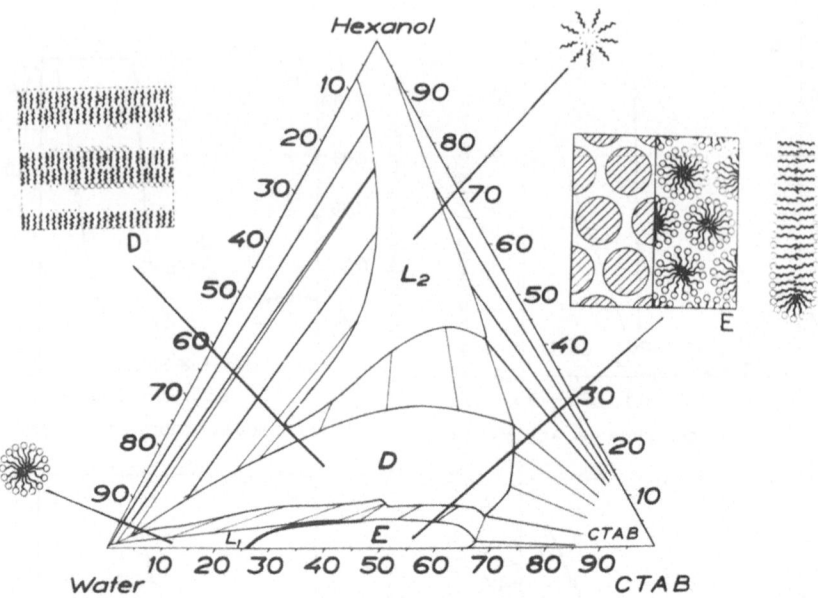

Fig. 2.20. Phase diagram (at 25 °C) from the work by Ekwall and co-workers (cf. Refs.[8, 86]) for the three-component system hexadecyltrimethylammonium bromide (CTAB) – hexanol – water. L_1 denotes a region with water-rich solutions; L_2 a region with hexanol-rich solutions; D and E are lamellar and hexagonal liquid crystalline phases, respectively. In the figure are also schematically indicated the structures of normal (L_1 region) and reversed (L_2) micelles as well as of the liquid crystalline phases. (From Ref.[95])

to the systematic studies of Ekwall, Fontell and Mandell[8, 86] detailed phase diagrams are available for a large number of systems and as an example we give in Fig. 2.20 the phase diagram for the system CTAB-hexanol-water, which well illustrates the qualitative behaviour of a large number of systems. It is thus a rather general observation both for cationic, anionic and nonionic systems, that a lamellar phase is formed as the solubilizate saturation limit is exceeded and also that this phase extends over a very large composition range. There is often a close correlation between the conditions in the liquid crystalline phase and the solubilization in the micellar solutions. It is suitable to describe solubilization in the three-component system as progressively varying but with changes with aggregation geometry. An inspection of a collection of phase diagrams shows that for a given surfactant the phase equilibria may be markedly different depending on the chemical structure of the solubilizate[8, 86]. This is well illustrated with sodium octanoate as surfactant. With long-chain alcohols, this surfactant gives extensive lamellar mesophase regions while for nonpolar solubilizates like p-xylene and octane, no lamellar phase is formed and with compounds such as octylnitrile, methyloctanoate, and octylaldehyde the lamellar phase region is small. The nonpolar solubilizates instead give rise to cubic mesophases. The phase denoted L_2 in Fig. 2.20 contains micelles of the reversed type; the formation of reversed micelles in particular depends on the organic compound and is absent for p-xylene and octane, for example. With long-chain alcohols there is an extensive formation of reversed micelles, on the other hand.

2.9 Some Special Aspects of Nonionic Systems

For nonionic systems, the intra- and intermicellar head group repulsions are much smaller than for ionic systems and this is expected to lead to important differences between the two classes of surfactants. One difference lies in the CMC values being almost 100 times smaller for nonionics than for ionics for the same alkyl chain length. We have, for example, $1.9 \cdot 10^{-4}$ M for dodecylglucoside and $0.87 \cdot 10^{-4}$ M for $C_{12}O(CH_2CH_2O)_6H$. For ionic surfactants, the CMC is strongly decreased in the presence of electrolyte while the effect is small for nonionic systems. The solubilization capacity is generally much bigger for nonionic systems than for ionic ones; this is exemplified by benzene solubilization[16] in Fig. 2.21.

A striking feature of nonionic systems is the clouding phenomenon. When the temperature of a micellar solution is raised the turbidity suddenly (at the cloud point) increases and at still higher temperatures a phase separation occurs. Concomitantly, the intrinsic viscosity shows a dramatic increase[96]. Most interesting is the observation that while the growth in size of micelles of ionic surfactants is accompanied by large rises in the ^{1}H NMR line width, no line width increases are observed in the region of the cloud point for nonionic surfactants[97].

The phase diagrams of two-component surfactant-water systems are typically quite different for nonionic and ionic compounds. As exemplified in Fig. 2.22 there are at low temperatures different liquid crystalline phases while at intermediate temperatures there may be a total mutual solubility of surfactant and water[98]. At higher temperatures, there is, as already noted, a separation into two phases with a very large two-phase region. One of the phases contains very little surfactant, while the other contains appreciable amounts of both components. The cloud-point curve can be described as a liquid-liquid solubility curve with a lower consolute tempera-

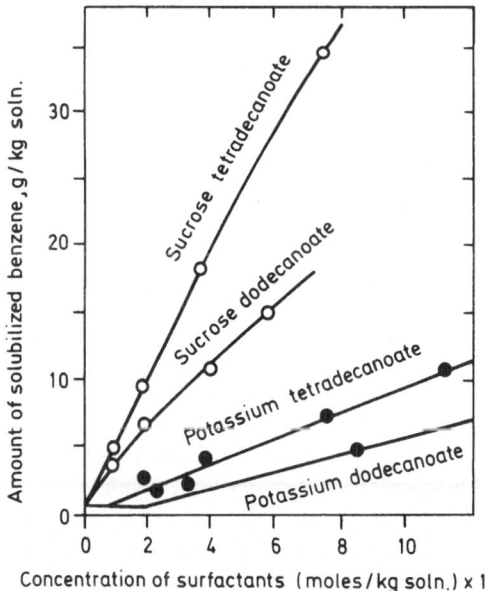

Fig. 2.21. Solubilization of benzene in surfactant solutions at 25 °C. (From Ref.[16])

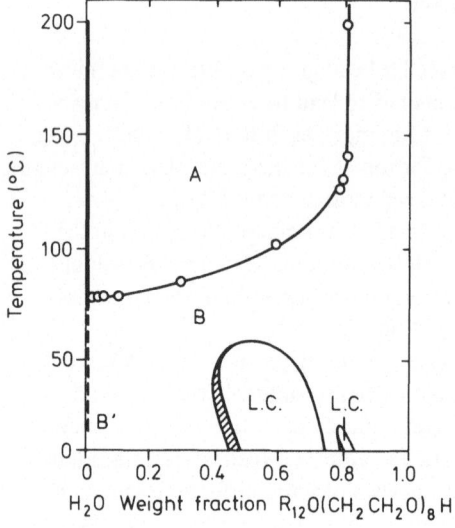

B. Lindman and H. Wennerström

Fig. 2.22. Phase diagram of the system $H_2O-C_{12}H_{25}O(CH_2CH_2O)_8H$ according to Shinoda[98]. *A* denotes a two phase area with two isotropic solutions in equilibrium, *B* a one phase area of an isotropic solution and LC liquid crystalline phases. (From Ref.[1])

ture. The surfactant saturation concentration above the cloud point corresponds approximately to the CMC. As for ionic surfactants, the phase diagrams of three-component systems nonionic surfactant-solubilizate-water may contain several mesophase regions with different structures. However, as is evident from the above, the appearance of the phase diagram may change extensively with temperature. Shinoda[99] has in particular determined partial phase diagrams for three-component systems where for example the surfactant content is kept constant. For similar weight percentages of water and solubilizate there is often a transition from a water-in-oil emulsion at low temperatures to an oil-in-water emulsion at high temperatures; this occurs in a narrow temperature range and is characterized by the so-called *phase inversion temperature* (PIT). In a narrow temperature range, there is formation of an additional isotropic phase, called the *surfactant phase.* The variation of the extension of the various one-phase regions with temperature is clearly demonstrated in phase equilibria studies, which also show that the surfactant phase has an independent phase region in a phase diagram for a fixed temperature[100].

2.10 Short-Chain Analogues of Surfactants and Other Nonmicelle-Forming Amphiphiles

The term micelle, as normally understood and as used in this work, refers to a large aggregate formed in a strongly cooperative association of amphiphilic compounds. A consequence of the cooperativity is that, as schematized in Fig. 2.23, the size distribution curve has a deep minimum; the appearance of a pronounced minimum in the size distribution curve is a good criterion for the formation of proper micelles. However, aggregation of amphiphilic compounds is of course not restricted to micelle formation and as indicated in Fig. 2.23 different types of size distribution curves

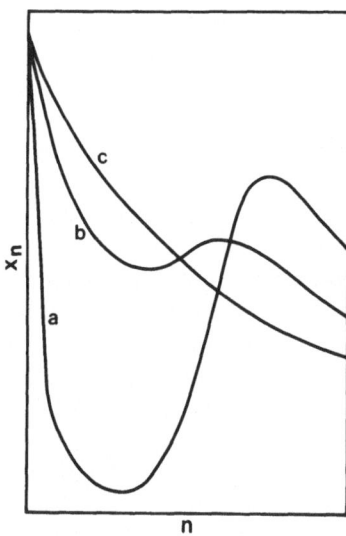

x_n

n

Fig. 2.23. Schematic picture of the aggregate size distribution well above the CMC for (a) a strongly cooperative association into micelles, (b) a weakly cooperative process, and (c) a noncooperative association

are then possible. Different amphiphiles behave rather differently and it may be useful to distinguish between the following three cases:

a) Aggregation to small aggregates.

b) Progressive noncooperative or weakly cooperative aggregation into large aggregates.

c) Aggregation effectively nonexistent in aqueous solutions but surfactant-type aggregation occurs in other phases.

An important approach for obtaining a good understanding of micelle formation is to study the effect of successively reducing the alkyl chain length. While surfactants with an n-alkyl chain with 8 or more carbon atoms are normally found to display typical micelle formation this is not the case for the lower homologues. A surfactant like sodium octanoate with 7 carbons in the alkyl chain is a borderline case with formation of appreciable amounts of both small aggregates and larger aggregates of 10–20 monomers[51, 101]. For sodium hexanoate, there appears to be no appreciable formation of larger aggregates. Studies of sodium salts of lower fatty acids have in particular been performed by Danielsson and co-workers[102, 103] using several different experimental techniques. Equilibria analyses in particular on the basis of potentiometric measurements have clearly demonstrated a formation of small aggregates for sodium alkanoates which do not form micelles. The formation of small aggregates prior to micelle formation is often referred to as *pre-micellar aggregation*. Although they can be postulated on theoretical grounds, it is very difficult to get experimental evidence for aggregates much smaller than micelles for surfactants with small CMCs. However, analyses of kinetic data have recently given evidence for the presence of small concentrations of oligomers[104].

The aggregation of bile salts is biologically interesting but has given rise to much confusion. While many authors have considered that micelle formation occurs and have presented CMC values, recent evidence shows that there is a progressive association extending over a wide concentration range with no critital concentration[105–108].

Bile salt association differs also in several other ways from micelle formation. For example, the counterion binding to the aggregates shows a marked increase over a narrow concentration range and the hydration varies in an irregular way[108]. It has been suggested[109] that there is a primary association into small aggregates due to hydrophobic interactions and a secondary aggregation of the primary aggregates by hydrogen-bonding.

The distinction between micellar and nonmicellar association may not always be immediately clear from experimental variable concentration data, but Mukerjee[6, 23, 24] has in instructive papers presented methods to distinguish between proper micelle fromation and other types of stepwise association. As he demonstrates, erroneous conclusions may easily be drawn and are abundant in the literature.

Swelling amphiphiles like lecithin and Aerosol OT do not form micelles in aqueous solution and therefore have a very low aqueous solubility. The reason is that due to nonpolar groups with large cross-sectional area, the spherical amphiphile aggregate is energetically disfavored in comparison with other geometries. That this is the case can be inferred from the extensive stability ranges of other aggregates, i.e., lamellar and reversed hexagonal mesophases. Furthermore, in the presence of a solubilizate and water, swelling amphiphiles have a marked tendency to form reversed micelles with a water core. An important feature of lecithin aggregation, which has important biological implications, is that bile acid salts very substantially increase the aqueous solubility of lecithin under the formation of very large mixed aggregates. Another conspicuous feature of lecithin is that a dispersion of its lamellar phase in water under ultrasonication gives clear solutions. These solutions contain so-called *vesicles*, often roughly spherical aggregates of a few hundred Ångström in diameter which consist of a single bilayer of lecithin separating an inner water core from the intervesicular aqueous solution. A quite similar structure is found in most cell membranes although here there is a large amount of protein as well as of cholesterol embedded in the phospholipid bilayer. Vesicles have attracted an enormous interest as convenient models of biological membranes and studies of molecular interactions, molecular dynamics and permeabilities of vesicles have been most important in enlarging our understanding of membranes.

3 Thermodynamics of Micelle Formation

3.1 Thermodynamic Models

The aggregation of amphiphilic molecules into micelles is, from a physico-chemical point of view, an example of the formation of a molecular complex. The total thermodynamic description of the aggregation would involve a series of stability constants, including their variation with salt concentration. In most applications, it is neither feasible to obtain such detailed information nor necessary from a practical point of view. The characteristic cooperative nature of the micellization makes it often possible to describe the aggregation process using only a few parameters. It has, for example, proven to be most useful to associate a *critical micelle concentration*, CMC, with each micelle forming amphiphile.

The CMC has its most clear-cut interpretation within the (pseudo) *phase separation model* of micelle formation. Although the micelles and the surrounding solution form a single phase, the amphiphile association shows a cooperativity that makes an analogy with a phase transition useful. Within this model, the CMC is the concentration at which the system enters a two phase region; the two pseudophases formed being the aqueous system and the micelles.

The phase separation model is particularly useful for describing the amount of micellized amphiphile and how molecular properties vary with amphiphile concentration. The average of a quantity Q (which can be a diffusion coefficient, a NMR chemical shift, a NMR relaxation time etc.) is determined by the fractions micellized, p_A^{mic}, and free, p_A^{aq}, amphiphile, so that for a total concentration C_{tot} larger than the CMC

$$<Q> = p_A^{mic} Q^{mic} + p_A^{aq} Q^{aq} = (1 - CMC/C_{tot})Q^{mic} + CMC\, Q^{aq}/C_{tot} \qquad (3.1)$$

where Q^{mic} and Q^{aq} are the values of Q in a micellar and an aqueous environment, respectively. Below the CMC $<Q> = Q^{aq}$. Thus by plotting $<Q> = Q_{obs}$ versus C_{tot}^{-1} one should obtain two straight lines which intersect at the CMC. Fig. 2.17 shows a typical application of this method. One can see that the intersection of the lines is well-defined but it is also evident that Eq. (3.1) only provides an approximate description of the experimental data. The model has two main deficiencies. As seen in Figs. 2.16 and 2.17 there is a smooth transition around the CMC instead of the predicted sharp kink. This is due to the fact that one does not have a proper phase transition. There is also sometimes a deviation from linearity at higher amphiphile concentrations.

An alternative to the phase separation model which has nearly the same simplicity is the *mass action law model*. In this model it is assumed that a single micellar species, of aggregation number n, is in equilibrium with the monomers:

$$n\, A_1 \rightleftarrows A_n; \quad [A_n]/[A_1]^n = K \qquad (3.2)$$

and

$$CMC \simeq K^{1/(n-1)} \qquad (3.3)$$

In reality several aggregation numbers of the micelles occur, but from the simple equilibrium in (3.2) one can make a number of relevant conclusions. The larger the value of n, the more cooperative is the association and the more one approaches phase separation behavior. This is illustrated in Fig. 3.1 which shows plots, for two values of n, of the fraction of the amphiphile that enters the micelle as a function of the total amphiphile concentrations. A comparison with the experimental data in Figs. 2.16 and 2.17 reveals that one can determine n from the smooth variation of Q_{obs} around the CMC. One can, however, note that the variation of Q_{obs} in the transition region is a measure of the cooperativity of the micellization, which can also be in-

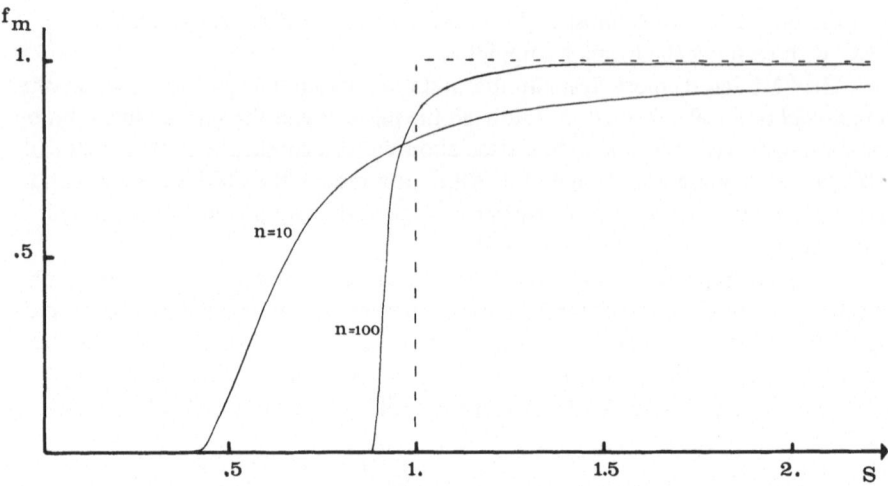

Fig. 3.1. The relative amount of amphiphile f_m that is incorporated into the micellar aggregate at an infinitesimal increase in the total concentration S as a function of total amphiphile concentration, $f_m = 1 - \dfrac{d[A_1]}{dS}$, calculated from the mass action law model (Eq. 3.2) with n = 10 and n = 100. The dotted line represents the behavior at a true phase separation (n = ∞)

fluenced by other factors than the aggregation number, such as the occurrence of premicellar aggregates.

To describe the micellar size distribution and how it is affected by the total amphiphile concentration more elaborate models are required. The phase separation model can be extended into the formally rigorous theory of *small system thermodynamics*[117, 118] developed by Hill. In this theory, the small system, in the present context the micelle, is regarded as immersed in a thermodynamic bath defining the value of the intensive variables. The consequence of having a finite system, relative to the infinite systems of classical thermodynamics, is that the number of independent intensive variables is increased by one. For a micellar system, the natural choice of independent variables is the temperature, the pressure and the chemical potential μ_{aq} of the monomeric amphiphile. The micelle size distribution function, F(n), is then a dependent variable that is determined by an equation of state, which at constant T and p gives a strict relation between F(n) and μ_{aq}. Naturally the form of this relation depends on the particular system, but the general formalism provides a number of relations between partial derivatives in much the same way as in classical thermodynamics. For a thorough description of the use of small system thermodynamics in describing micellar systems, the reader is referred to the last chapter of Hill's book[118] and to a review article by Hall and Pethica[119].

In most cases when the micelle size distribution has been analyzed quantitatively the *multiple equilibrium model* has been used. This model can be formulated either, in analogy with Eq. (3.2), as a number of equilibria

$$n A_1 \rightleftarrows A_n \qquad n = 2, 3 \dots \qquad (3.4)$$

with equilibrium constants

$$K_n = [A_n]/[A_1]^n \qquad (3.5)$$

or as a stepwise aggregation

$$A_1 + A_{n-1} \rightleftarrows A_n \qquad n = 2, 3 \ldots \qquad (3.6)$$

where the equilibrium constants are

$$K'_n = [A_n]/\{[A_{n-1}][A_1]\}. \qquad (3.7)$$

The schemes (3.4) and (3.6) are thermodynamically equivalent while they have different implications for the kinetic behavior (cf. Sect. 5). The equilibrium constants are related by

$$K_n = \prod_{i=2}^{n} K'_i. \qquad (3.8)$$

It is sometimes useful to express the equilibria in terms of chemical potentials

$$-RT \ln K_n = n(\mu_n^\theta - \mu_{aq}^\theta) \qquad (3.9)$$

where μ_n^θ is the standard chemical potential per monomer in a micelle of aggregation number n. Both K_n and μ^θ depend on the units used for expressing concentrations. When these are chosen as mole fractions of amphiphilic molecules in the aqueous region, X_1, and in aggregate n, X_n, Eqs. (3.5) and (3.9) imply

$$X_n = n\{X_1 \exp[(\mu_{aq}^\theta - \mu_n^\theta)/RT]\}^n \qquad (3.10)$$

The total mole fraction of amphiphile, S is

$$S = \sum_i X_i \qquad (3.11)$$

and the value of $(\mu_{aq}^\theta - \mu_n^\theta)$ for all n defines the micelle distribution for a given S. In practice one cannot extract information on all μ_n^θ, but even the qualitative features of μ_n^θ can be of interest. For example, as discussed in Sect. 3.4, the asymptotic behavior of μ_n^θ determines how the system behaves at high amphiphile concentrations.

In the previous discussion, activity coefficients have been totally neglected, for simplicity, but they should be included in a proper treatment. However, for ionic amphiphiles, where activity corrections are expected to be most important, additional complications arise. If the counterions are explicitly incorporated in the equilibria the number of possible chemical species is greatly enhanced making a detailed analysis even more complex. Furthermore a description of a process in terms of an equilibrium constant is only really suitable when the forces involved are of a short range nature. The electrostatic interactions in an ionic micellar system are of a long-range nature and seem to be better discussed on the basis of electrostatic equations as discussed in Sect. 6.

3.2 The Hydrophobic Interaction

The formation of amphiphilic aggregates can be seen as a compromise between the tendency of alkyl chains to avoid contact with water and the strong affinity of the polar groups to water. A thermodynamic description of micellar systems should thus include both these aspects. The electrostatic effects are discussed in Sect. 6 so we will concentrate here on the interactions involving the nonpolar part of the amphiphile.

The low solubility of hydrocarbons and other mainly apolar substances in water has been ascribed phenomenologically to the hydrophobic interaction. The hydrophobic free energy can be defined[4] as the difference between the standard chemical potentials of an apolar solute at infinite dilution in a hydrocarbon solvent μ_{HC}^θ and in water μ_{aq}^θ

$$\mu_{HF}^\theta = \mu_{HC}^\theta - \mu_{aq}^\theta. \tag{3.12}$$

This represents a thermodynamical approach and one does not assume anything about the molecular origin of the difference in free energy. A circumstance that makes the introduction of μ_{HF}^θ useful is that in a majority of cases there is a constant increment of μ_{HF}^θ in a homologous series. The increment is furthermore roughly independent of the functional group so that similar values apply to alcohols, carboxylic acids and pure hydrocarbons[4, 120–122]. Thus for an homologous series of compounds

$$\mu_{HF}^\theta = (a - bn_C)\, RT \tag{3.13}$$

where a and b are constants and n_C is the number of carbon atoms. Table 3.1 shows values for a and b for a number of series of compounds and one can see that the constant b is around 1.4–1.5.

Table 3.1. Values of the constants a and b in the relation $\mu_{HF}^\theta/(RT) =$ a–b n_C for the hydrophobic free energy of an homologous series of compounds. (T = 298 K)

Compounds	a	b
$CH_3(CH_2)_{n-2}CH_3$[a]	− 4.11	1.49
$CH_2 = CH-(CH_2)_{n-3}CH_3$[a]	− 2.54	1.49
$CH_2 = CH-(CH_2)_{n-4}-CH = CH_2$[a]	− 1.52	1.49
$CH_3(CH_2)_{n-1}OH$[b]	1.40	1.39
$CH_3(CH_2)_{n-2}COOH$[c]	7.18	1.39

a From Ref. [121];   b From Ref. [122];   c From Ref. [120].

There are several additional ways of empirically correlating the hydrophobic free energy with structural parameters. There is a good linear relation between the surface area of the apolar solute and μ_{HF}^θ[123, 124]. In this scheme one directly accounts for the fact that the magnitude of μ_{HF}^θ is smaller for branched than for normal alkanes. Another suggestion is [125] that it is the number of hydrogens in the apolar part that determines μ_{HF}^θ. The virtue of this approach is that one correctly describes the difference in μ_{HF}^θ between saturated and unsaturated hydrocarbon chains.

A further insight into the nature of the hydrophobic interaction is obtained by separating the free energy into enthalpic and entropic parts

$$\mu_{HF}^\theta = h_{HF}^\theta - T\, s_{HF}^\theta \qquad (3.14)$$

A characteristic feature of the hydrophobic interaction is that it is dominated by entropy effects. Both the temperature dependence of alkane solubilities in water[126, 127] and direct calorimetric measurements[128] show that h_{HF}^θ is close to zero at room temperature. Some calorimetric data for heats of solution of hydrocarbons in water are shown in Table 3.2. A further noticeable feature is that h_{HF}^θ is temperature dependent due to the rather large heat capacity, c_p^{HF}, associated with the hydrophobic interaction. From a systematic calorimetric study of a series of compounds with rather short alkyl chains[129] it was found that

$$c_p^{HF} \simeq 90\ J\ mol^{-1}\ K^{-1} \qquad (3.15)$$

per CH_2-group. It was also concluded that c_p^{HF} seems independent of temperature. On the basis of the regular behavior of the heat capacity, Gill and Wadsö[125] were able to summarize the thermodynamic data for the hydrophobic interaction of hydrocarbons as

$$h_{HF}^\theta(T) = c_p^{HF}(T - T*) \qquad (3.16)$$

$$s_{HF}^\theta(T) = -\frac{\mu_{HF}^\theta(T*)}{T*} + c_p^{HF} \ln T/T* \qquad (3.17)$$

Table 3.2. Calorimetric parameters for the dissolution of hydrocarbons in water (T = 298 K) (From Ref.[125])

Compound	$\Delta H^\theta (kJ\ mol^{-1})$	$\Delta C_p^\theta (J\ K^{-1} mol^{-1})$	T* (K)[a]
Benzene	2.08	225	289
Toluene	1.73	263	292
Ethyl benzene	2.02	318	292
Propyl benzene	2.3	391	292
Pentane	− 2.0	400	303
Cyclohexane	− 0.1	360	298
Hexane	± 0.0	440	298

[a] T* is the temperature where $\Delta H^\theta = 0$

where $T*$ is the temperature at which h_{HF}^{θ} is zero. A mean value for seven hydro-carbons gave $T* = 295$ K and the hydrophobic free energy for a number of compounds could be summarized as

$$h_{HF}^{\theta}(T) = T(6.3 + 11.6 \, n_H)/295$$
$$- 0.033 \, T \, n_H \, (\ln T/295 + 295 \, T) \, kJ \, mol^{-1} \tag{3.18}$$

where n_H is the number of hydrogens in the hydrocarbon chain.

The equations describing the thermodynamical parameters do not directly give information on the molecular nature of the hydrophobic interaction. The general nature of the effect has, however, nontrivial implications. Thus the solubilities of methane and noble gases behave according to the hydrophobic effect and there is no qualitative difference between the behavior of benzene and hexane. This indicates that the hydrophobic interaction is not primarily due to the chemical properties of the solute but rather a property of the solvent water. This conclusion is corroborated by a number of studies of the physical properties of aqueous solutions of apolar solutes. The partial molar volumes of hydrocarbons in water is anomalously small[130, 131], the average rotational correlation time of the water molecules is long, [132] etc. These effects can be interpreted in terms of a hydrophobic hydration of the apolar solute. The water molecules close to the solute form a stronger hydrogen bonded network than in bulk water, and the solute occupies interstitial positions in the open structure. The solution structure is thus analogous to the one found for crystalline clathrates of, for example, noble gases and alkyl ammonium salts[133]. However, for concentrated solutions of tetrabutylammonium fluoride no indications of a clathrate-like structure were found from the X-ray diffraction pattern[134]. For a thorough discussion of hydrophobic hydration the reader should consult articles in the monograph on water edited by Franks[135].

The outlined model of hydrophobic hydration gives also a rationalization of the observation that μ_{HF}^{θ} is dominated by entropy effects. The enthalpy increase associated with the creation of a cavity to be occupied by the solute is compensated by an increased hydrogen bonding to the remaining water molecules, while the increased solvent structure gives a decrease in entropy. Recent theoretical studies using analytical statistical mechanical techniques[136, 137] as well as Monte Carlo simulations[138, 139] have confirmed these conclusions. The Monte Carlo calculations show explicitly that the enthalpy gain in an increased water-water interaction can more than compensate for the creation of a small cavity.

In aqueous solutions of amphiphiles, the hydrophobic interaction will provide a thermodynamic driving force for an aggregation process. If the forces involving the polar groups, which oppose the aggregation, are independent of the carbon chain length one expects a regular dependence of the CMC on alkyl chain length. According to Eqs. (3.3), (3.9) and (3.13)

$$\ln CMC \simeq \frac{1}{n} \ln K_n \simeq const + \mu_{HF}^{\theta}/(RT) \simeq a_1 - b_1 \, n_C \tag{3.19}$$

where a_1 and b_1 are constants. Experimentally one finds indeed a linear relation between $\ln CMC$ and the number of carbons n_C in the alkyl chain (cf. Fig. 2.3). For nonionic amphiphiles[4, 113] $b_1 \simeq 1.2$ (T = 298 K) which is significantly less than the value 1.5 found from the solubility data for hydrocarbons. The origin of this discrepancy is not well understood. One possibility is that the anchoring of the polar head at the micellar surface restricts the alkyl chain motions sufficiently to give a sizable decrease in entropy. For ionic amphiphiles the value of b_1 is even smaller[4]. This is mainly caused by ion binding effects and by measuring the CMC in solutions with an excess of salt the coefficient b_1 is increased to the value found for nonionic amphiphiles. These regularities seem to establish that the main driving force for the formation of micelles is the hydrophobic interaction.

3.3 The Enthalpy of Micelle Formation

The enthalpy changes involved in a micelle formation process can be obtained experimentally using several different techniques. From the temperature dependence of the CMC, the enthalpy is obtained through a van't Hoff type relation

$$RT^2 \frac{d}{dT} \ln CMC = -\Delta H^\theta \tag{3.20}$$

Enthalpy changes can also be measured directly in a calorimeter. The temperature dependence of kinetic parameters can be interpreted in terms of ΔH values. In analyzing the enthalpies it is essential to recognize the chemical processes that actually contribute to the particular ΔH-value.

In the same way as in the discussion of equilibria one can regard the total process

$$n A_1 \rightarrow A_n; \quad n \Delta H_n \tag{3.21}$$

or one can distinguish between the different steps

$$A_1 + A_{n-1} \rightarrow A_n; \quad \Delta H_n' \tag{3.22}$$

where the enthalpies are related through

$$n \Delta H_n = \sum_{i=2}^{n} \Delta H_i' \tag{3.23}$$

Within the phase separation and mass action law models, an application of Eq. (3.20) would yield ΔH_n, while the individual $\Delta H_i'$-values have no meaning in these models. However, these two models are too simplified in the present context and one has to recognize that there is a distribution of micellar sizes and that this distribution is temperature dependent. Thus when measuring the CMC at different temperatures, the aggregation is a somewhat varying process and Eq. (3.20) is not strictly applicable[140]. The analysis of ΔH values obtained from Eq. (3.20) has recently been thoroughly discussed by Muller[31]. He concludes that one obtains reasonably accurate values for ΔH_n but that $\Delta H_i'$ for i close to the optimal aggregation number can be substantially different.

Experimentally the CMC is usually weakly temperature dependent[12] (cf. Fig. 2.5) indicating that ΔH_n is close to zero. For the cases[141-147] where calorimetric determinations have been performed this conclusion has been confirmed.

An interesting aspect of the enthalpies of micellization is obtained from kinetic measurements. As described in Sect. 5, one observes two kinetic relaxation times and from the temperature dependence of the fast process one can deduce values for $\Delta H_i'$ for i around the optimal micelle size. The slow process, on the other hand, involves the formation of rare pre-micellar aggregates and from the temperature dependence of this process one obtains $\Delta H_i'$ for these intermediate aggregates[104]. That ΔH_n is close to zero would then be explained by the fact that $\Delta H_i'$ changes sign as i varies so that there is a cancellation effect in the sum in Eq. (3.23). It is not straightforward to find the molecular origin of the i-dependence of $\Delta H_i'$. As discussed in Sect. 3.2 the hydrophobic free energies are mainly of entropic origin. Thus it seems reasonable to ascribe the variation in $\Delta H_i'$ to a variation in interactions involving the polar head-groups. One appealing possibility is that an aggregate size dependence of the counterion binding accounts for the observed effect. This could be in accordance with an ion condensation behavior which is further discussed in Sect. 6.

In calorimetric studies of micelle formation it is often difficult to relate the measured enthalpy changes to specified steps in the aggregation process. Instead one perferably determines the partial molar enthalpy h_A of the amphiphile as a function of concentration[12]. The ideal case of the phase separation model predicts that h_A is constant up to the CMC where it discontinuously jumps to another constant value. The behavior of h_A in real micellar systems is more complex as seen in Fig. 2.12. Similar data have been obtained for several other amphiphiles[148, 149]. The deviations in h_A from the standard value at infinite dilution appear clearly below the CMC, but at these concentrations one has a compensating change in the partial molar entropy. This effect might be due to a repulsive interaction between the hydrophobically hydrated alkyl chains leading to a breakdown of the water structure with a concomitant increase in entropy.

3.4 Model Calculations of the Free Energy of Micelle Formation. Micelle Size and Shape

The first attempt to theoretically describe the formation of micelles in terms of molecular interactions was made by Debye[150,151]. He considered a hydrophobic and an electrostatic energy contribution and saw the formation of micelles as due to the competition of these two forces. Although the expressions for the free energies were qualitatively incorrect, in the sense that the amphiphile aggregation was predicted to be anticooperative instead of cooperative[152], Debye correctly identified the important interactions. Hoeve and Benson[153] attempted a detailed statistical mechanical description of the micelles and the monomers, but too little was known at that stage about the molecular interactions to allow any quantitative predictions. The formal statistical mechanical approach was extended by Poland and Scheraga[154-156] but although progress had been made with regard to the hydrophobic interaction[157] the quantitative results were of limited value.

In more recent attempts[158−167] to quantitatively describe the micellar aggregation, empirical expressions for the hydrophobic energy like the one in Eq. (3.13)
have been used. It is furthermore recognized that even when the micelles have formed
there is a nonnegligible contact between apolar groups at the micellar surface and the
water giving a positive contribution to the free energy of micelle formation. The most
problematic part in estimating the total free energy has been the contributions from
the polar groups. Tanford uses an empirical expression obtained from surface pressure data while others use values obtained from the Debye-Hückel[162−164] or the
Gouy-Chapman[167] theories. The electrostatic energy has also been approximated
by that of a capacitor[165] or simply of a charged sphere[166].

The calculated dependence of μ_n^θ on the aggregation number n gives the possibility of following the aggregation process in detail. These studies indeed show that the
formation of micelles is a cooperative process[163, 164]. It appears that the source of
the cooperativity is analogous to the one which leads to the formation of a separate
liquid phase for ordinary hydrocarbons. In the micellar system the cooperativity is
partly broken when a spherical aggregate is formed so that the polar groups cover the
micellar surface and a further increase in size must lead to change in the aggregate
shape. From the calculated values of μ_n^θ one derives values of the CMC which are in
good agreement with experiments. However, it is not clear if the variation of μ_n^θ for
n-values larger than those of a spherical aggregate is described correctly. An indication of the difficulties is obtained if one compares what the different formulations
predict about the preferred aggregate shape. Tanford[160] concludes that an oblate
micelle is the most favored shape. This is in conflict with experimental studies in
which one almost exclusively[34, 35, 83, 97, 167−170] finds that the micelles attain a
rodshaped structure if they grow in size. Ruckenstein and Nagarajan find similar-
ily[164] that for a soap with a single alkyl chain a vesicle aggregate is of similar stability as a micelle. Isrealachvili et al.[165] find that a toroidal structure is the most favorable one. In our opinion, one source of these shortcomings of the model calculations
is that the electrostatic term has not been considered properly. It is, for example,
more unfavorable to assemble charged amphiphiles in large compact aggregates than
in small ones.

As is clear from the discussion of micellar shape this property is closely related
to the micelle size. Israelachvili et al.[165] made the important observation that the
growth of micelles at high amphiphile concentrations is determined by the asymptotic behavior of μ_n^θ as n increases. The growth of an aggregate in one dimension
only, like a rod, is non-cooperative. This leads to a continuous increase in aggregate
size and a transition to a liquid crystalline structure, with infinite aggregates, occurs
first when the inter-aggregate interactions become important. A lamellar-like aggregate grows in two dimensions and this is a cooperative process, which makes it
possible for a phase change to occur even in the absence of inter-aggregate interactions[165, 171].

3.5 Temperature Dependence of the Micelle Size

For ionic amphiphiles it seems to be a general trend that, although the CMC is rather
insensitive to temperature changes, the tendency to form micelles larger than spher-

ical ones increases with decreasing temperature[33, 172]. This conclusion is consistent with the finding that $\Delta H'_n$ of Eq. (3.22) is negative for aggregation numbers n typical of the spherical micelle. Thus the addition of a monomer to an already formed micelle is exothermic and the enthalpic contribution becomes more important the lower the temperature. The forces that determine the growth of the micelles are not so much associated with the hydrophobic interaction, since in the spherical micelle water-hydrocarbon contact has already been substantially reduced. Instead it is the interplay between the repulsion of the polar groups and the possible packing of the alkyl chains within the aggregates[4, 173] that determine the preferred aggregate shape.

For nonionic amphiphiles it is usually stated[2, 12, 113] that the micelle size increases with increasing temperature. The conclusion is mainly based on light scattering data. However, recent studies[97, 174] indicate that secondary aggregation of spherical micelles occurs when the temperature is increased and it seems to be a serious possibility that the light scattering data has been misinterpreted. The intense light scattering at the cloud-point is probably not due to large aggregates but rather due to the large fluctuations in the local micelle concentration that occur when the phase boundary is approached. This mechanism is analogous to the strong light scattering due to density fluctuations observed for simple fluids close to the critical point[175].

The occurrence of two phases at the higher temperature shows that there is a net attraction between the micelles. Since the clouding is a general phenomenon it appears that the attraction is not due to specific effects but could be caused by van der Waals' forces. This interaction is temperature independent and one has also to invoke a temperature dependent repulsive force which might be due to the repulsion between water molecules in the solvation shells of the micelles[137, 176].

3.6 Thermodynamic Aspects of Mixed Micelles and Solubilization

The preceding discussion has been confined to two-component systems, amphiphile-water. In a large number of cases of practical importance one adds one (or more) additional component(s). Depending on the nature of the additive one can recognize different effects. If it is an amphiphile it is usually found that the micelles which form in the solution are of mixed composition. Under the assumption that the amphiphiles mix ideally in the micellar aggregate, Shinoda[177] has derived expressions for the effective CMC of the amphiphile mixture. For nonionics

$$CMC = \sum_i X_i (CMC)_i^2 / \sum X_i (CMC)_i \qquad (3.24)$$

where X_i is the relative amount of amphiphile i and $(CMC)_i$ its critical micelle concentration. For ionic amphiphiles the electrostatic effects make the equation equivalent to Eq. (3.24) somewhat more complex. The assumption of ideal mixing seems to apply reasonably well for amphiphiles of similar structure. An interesting exception is mixtures of hydrocarbon and fluorocarbon amphiphiles where the deviations from ideality are so large that one might suspect that separate hydrocarbon and fluorocarbon micelles form in the solution[25, 26]. Mixtures of nonionic and ionic amphiphiles are also nonideal but it seems clear that mixed micelles are actually formed[178]. This observation supports the view that the basic driving force for the formation of micelles is the same for ionic and nonionic amphiphiles. A special type of mixed micelle is found in the physiologically important system water-sodium cholate-lecithin[109, 179]. Both cholate and lecithin are amphiphilic but they do not form micelles in pure form, but when mixed typical micellar aggregates appear, on the other hand. The mixed system is thus clearly nonideal and one has probably a rather specific arrangement between the two components in the micelles. A further type of nonideal mixed micelle is obtained in mixtures of anionic and cationic amphiphiles. At equimolar concentrations neutral micelles can form and the CMC is drastically decreased. There is also an increased tendency to form rod-shaped micelles[180].

When a semipolar, e.g., an alcohol, or a nonpolar, e.g., an alkane, substance is added to a micellar system the additives are solubilized in the micelles. The presence of a solubilized molecule in the micelle reduces the activity of the amphiphile in the aggregate. It is thus a natural consequence that the CMC (with respect to amphiphile concentration) is lowered. A strict thermodynamic analysis[4, 181, 182] gives, in analogy with Henry's law:

$$CMC(X_A) = CMC\ (X_A = 0) - KX_A \qquad (3.25)$$

where X_A is the mole fraction of the additive. This relation has, for example, been experimentally verified with alkyl alcohols as the additive[2] (cf. Fig. 2.7).

Measurements of thermodynamic parameters can provide important information on the chemical nature of the solubilization process. The partial molar volumes of a hydrocarbon in a micellar solution is very similar to the value obtained for a hydrocarbon phase, and differs significantly from that obtained in water[39, 183]. This gives strong support to the notion that the interior of the micelles is hydrocarbon-like. Calorimetric measurements on the solubilization process have been largely confined to semipolar additives. The general picture is that these are solubilized with their polar group at the micellar surface and with the apolar part towards the micellar interior. From such a picture one expects that the heat of solution in the micellar system is different from that found with pure water or pure hydrocarbon as solvent. This is indeed generally found[184, 185] showing the specific nature of the solubilization process.

4 Micelle Structure

4.1 The Association Process

The strong cooperativity of amphiphile association into micelles is well established and for long-chain surfactants it is often a good approximation to consider micelle formation as analogous to a phase separation. Even if the concentration dependence of many physico-chemical properties, within experimental accuracy, is in concordance with a phase separation model, certain properties, as well as data for short-chain micelle-forming surfactants, do show that there is a consecutive association. A full equilibrium analysis (including the determination of all the equilibrium constants) without simplifying assumptions is evidently out of the question for typical aggregation numbers around 50 or above, especially since the various aggregates have very similar intrinsic spectroscopic properties. Even so a large number of important questions regarding the association process can now be answered: e.g., what is the general shape of the size distribution curve, how large are the deviations from the phase separation model for different cases, what are the concentrations of the rare aggregates of intermediate size, and how does the surfactant monomer concentration change with the total surfactant concentration?

Because of the strong cooperativity, it is clearly very difficult to monitor the formation of aggregates considerably smaller than the abundant micelles. Thus the aggregation into small aggregates is dominating only in an extremely narrow concentration region. One approach to study the initial aggregation is to lower the cooperativity by reducing the hydrophobic character and working with short-chain analogues of surfactants. Such studies have clearly demonstrated the formation of small aggregates for the lower sodium alkanoates from butyrate and upwards[38, 72, 101–103]. For example, for sodium n-pentanoate and n-hexanoate there is a formation of appreciable amounts of tetramers as well as of somewhat larger aggregates, and for octanoate there is a formation of small aggregates with 5 monomers and larger ones with between 9 and 17 monomers. A limitation of these studies is that they require the addition of high electrolyte concentrations to swamp variations in activity coefficients. By using spectroscopic methods, such as ^{13}C and ^{19}F NMR spectroscopy, this can be avoided. A ^{13}C NMR chemical shift study[51] of aqueous solutions of sodium n-hexanoate and sodium n-octanoate shows that there is in both cases an appreciable formation of small aggregates (trimers, tetramers). For hexanoate, there seems to be no appreciable formation of large aggregates, while for octanoate the data can be reconciled with the formation of larger aggregates (ca. 10–11 monomers). In neither case, can the aggregation be said to lead to proper micelles although sodium octanoate appears to be a borderline case.

On the basis of the phase separation model, it is straightforward to predict the concentration dependence of a given measured property. For example, a plot of the NMR chemical shift of the amphiphile, or the amphiphile self-diffusion coefficient, versus the inverse surfactant concentration should in the ideal case consist of two straight-line segments intersecting at the CMC [Eq. (3.1)]. In practice a considerable curvature is observed for not too long alkyl chains as exemplified by the ^{13}C chemical shifts of solutions of nonylammonium bromide in Fig. 2.16. Lengthening of the alkyl chain leads to a considerable reduction of the curvature as can be seen for example in the ^{23}Na NMR chemical shift studies[59] of solutions of several sodium alkylsulfates (Fig. 2.17). The variation of the amount of micellized dodecyl-sulfate with concentration according to conductivity and solubilization studies is depicted in Fig. 4.1.

From the strong cooperativity and from the approximate invariance of the mean aggregation number to concentration it can be inferred that intermediate size aggregates are present in very low concentrations and this is well documented in kinetic

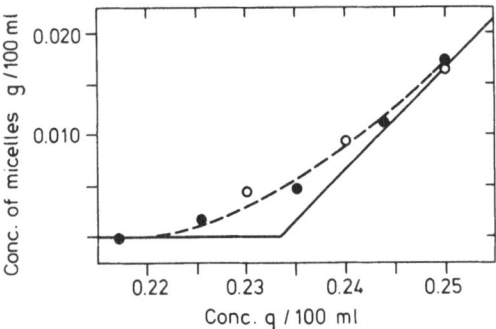

Fig. 4.1. Approximate micellar concentration around the CMC for sodium dodecylsulfate solutions estimated from conductivity and dye solubilization. (From Ref.[186])

studies[104, 187, 188] (see Sect. 5). It has been possible from studies at variable concentration of added electrolyte to obtain an average aggregation number of ca. 7 at the minimum of the size distribution curve for the case of sodium tetradecylsulfate and dodecylsulfate[104]. It has also been demonstrated that the intermediate aggregates may have concentrations which are several orders of magnitude below those of abundant micelles.

If one considers solely the consecutive equilibria, the concentration of monomer can only increase with increasing total amphiphile concentration even above the CMC. (Apart from the trivial decrease in the monomer concentration calculated on the total volume which may arise when the micelles occupy a substantial volume fraction). However, if one realizes that micelles are not only composed of amphiphile, the result may be different. Thus counterion binding helps to stabilize the micelles and for ionic surfactants it can be predicted that the monomer activity may decrease with increasing surfactant concentration above the CMC. Good evidence for a decreasing monomer concentration above the CMC has been provided in the kinetic investigations of Aniansson et al.[104], and recently Cutler et al.[46] demonstrated, from amphiphile specific electrode studies, that the activity of dodecylsulfate ions decreases quite appreciably above the CMC for sodium dodecylsulfate solutions (Fig. 2.14).

4.2 Micelle Size and Shape. Polydispersity

Classic light scattering has been the main source of information on micelle size but also viscosity, tracer diffusion and quasi-elastic light scattering are frequently applied methods[2, 4, 9, 33]. However, it seems that in one respect or another some criticism may be raised of these methods. Light scattering is strongly affected by electrostatic intermicellar interactions requiring extrapolation to low concentrations or addition of electrolytes. (Except for spherical micelles, the micelle size depends strongly on concentration ant then the extrapolation procedure seems to have no sound basis). Even so, the method may be subject to error as can be inferred from the probable misinterpretation of light scattering data of dilute nonionic systems (see Sect. 3). Tracer diffusion studies are much less affected by intermicellar interactions but require either the addition of a solubilizate[111] which, although present in low concentration, may affect aggregation or, if surfactant diffusion is monitored[37], information on the association process. In quasi-elastic light scattering (QLS)[175] one has the somewhat similar problems as in tracer diffusion studies with intermicellar interactions but additionally it has been argued[189-191] that a mutual diffusion coefficient is measured. The conversion of this into a self-diffusion coefficient may become cumbersome at high surfactant concentrations. Berne and Pecora[175] give most appropriate aspects of this problem. Thus because of concentration fluctuations, the normal spherically symmetric configuration of counterions ions around a spherical micelle can become asymmetric, so that the micelle experiences a nonzero electrical force. This instantaneous force due to the counterion distribution accelerates the micelle motion so that it becomes larger than would be expected on the basis of Stokes' law. This effect is expected to be larger for dilute solutions because of large relative fluctuations. The problems encountered with the quasielastic light scattering method can, however, be largely overcome by using high concentrations of added salt as shown by Mazer et al.[33].

In the present authors' opinion, some of the difficulties have often not been adequately considered and, therefore, the vast material on micellar aggregation numbers given in the literature may be subject to error; results are probably in most cases qualitatively correct but as suggested by the interpretations for nonionic systems even this is not certain. If one considers the mass of ionic surfactant micelles, one finds for many cases that the aggregation numbers are close to those expected for spherical aggregates with a radius corresponding to the length of the surfactant with an extended hydrocarbon chain. The general picture seems to apply in most cases for conditions of low amphiphile concentrations and in the absence of added electrolyte.

The methods mentioned above do not measure the micellar aggregation number itself but rather some micellar size. A direct determination of aggregation numbers can be performed by analysing some physico-chemical parameter in terms of the equilibria involved; equilibrium analyses on the basis of potentiometric data have been pioneered by Danielsson and co-workers who studied short-chain, not typically micelle-forming, amphiphiles in the presence of added electrolyte (see above). ^{13}C NMR chemical shift studies have recently been performed with the same purpose. For nonylammonium bromide, data (Fig. 2.16) for the different carbons were in excellent agreement with a simple model considering aggregation into a single micelle size (aggregation number 35 ± 2). One interesting observation was that the error square sum increases much faster when the aggregation number is decreased from the optimum value than when it is increased from that value. This study indicates that the distribution of micelle size is narrow and somewhat unsymmetrical around the average value.

As expected, micelle size increases with increasing alkyl chain length of the amphiphile; in several cases, the growth in size is small and corresponds to a retention of the same micellar shape, in others, shape changes must be invoked to explain the data. A significant observation is that although the CMC depends little on headgroup structure and counterion, micelle size can vary by orders of magnitude. Electrolyte addition to ionic systems in general gives an increased micelle size; in most cases the effect is rather small while in others dramatic changes occur. There seems for all systems to be a decrease in micellar size with increasing temperature; the change is rather small for most cases, but for certain cases enormous variations may occur over a narrow range. In some instances, very pronounced increases in micellar size with increasing surfactant concentration are observed while in most cases the effect is very small; while some increase is expected in all cases it is often difficult to get conclusive evidence since intermicellar interactions in most methods give apparent increases that are difficult to correct for. From the above, it can be seen that, in contrast to the CMC, micellar size varies with various factors in a manner which is complex and presently difficult to predict. We shall, therefore, not attempt to give a unified picture of the matter but rather present a few examples.

For hexadecyltrimethylammonium bromide (CTAB) there is evidence of several types (X-ray diffraction[83], conductance anisotropy[192], flow birefringence[193], ^{1}H[194], ^{2}H[67], ^{14}N[195] and ^{81}Br[58] NMR, linear dichroism[196] and viscosity[35] (Fig. 2.9) that there is a transition from closely spherical to very long rod-like aggregates at a concentration of 0.2–0.3 M. Significant is the observation of ^{2}H quadrupole splittings similar to those of the hexagonal mesophase[67]. This system well exemplifies the great sensitivity of micellar size to a number of parameters. Increasing the temperature from 30 to 50 °C, as well as substituting Cl$^-$ for Br$^-$ as counterion, eliminates the transition[35, 58]. Addition of small amounts of simple solubilizates such as benzene or a long-chain alcohol (hexanol, octanol etc.) may markedly facilitate the transition to rod-like micelles while alkanes have no effect[56, 58, 197]. Decreasing the alkyl chain length considerably increases the transition concentration or eliminates the transition completely[58]. Viscoelasticity is produced with certain organic counterions like salicylate or by certain solubilizates like α-methylnaphtalene[29, 198]. The micellar shape and interactions giving viscoelasticity, which depends on subtle effects in counterion or solubilizate structure, are not yet completely unterstood but it seems that long-range electrostatic intermicellar interactions are important[172, 199].

Mukerjee[200, 201] has discussed the relation between micellar size, polydispersity, the sphere-to-rod transition and the equilibrium constants of the step-wise association. For many systems, the degree of polydispersity is low as shown by the close equality of the number average and the weight average degrees of association. Such a behavior is predicted if after the increase in association constant up to a certain aggregation number there is a region of marked anti-cooperativity with the equilibrium constant decreasing with aggregation number. For micellar systems showing the sphere-to-rod transition, this anti-cooperativity, which lies in head-group repulsions, is partly eliminated. A step-wise association model was derived for the case of large micelles which predicts that the weight average aggregation number should be twice the number average value and also that the former should increase in proportion to the square root of the concentration of micellized surfactant[200]; this is in agreement with some experimental data. As shown by the case of CTAB, the elimination of head-group repulsions can be brought about by counterions which may approach the charged groups closely or intercalate between them or by certain solubilizates which are located in the head-group region. For the cases of growth to very long rod-like micelles, the polydispersity becomes large. While the formation of rod-like micelles is well demonstrated for several cases, other geometries of the large nonspherical micelles have not been found which is in agreement with theoretical predictions[165, 171].

Detailed studies of micellar size and shape using quasi-elastic light scattering have recently been presented for sodium dodecylsulfate[33]. To reduce the influence of intermicellar interactions, the solutions contained rather high concentrations of NaCl (0.15–0.6 M). The measured diffusion coefficients were assumed to approximate the self-diffusion coefficients which were converted into micellar radii and aggregation numbers using the simple Stokes-Einstein relation which neglects intermicellar interactions which may be reasonable at these high electrolyte concentrations. The aggregation number was found to be ca. 60 under conditions of high temperature or low salt concentration. Decreasing temperature, increasing NaCl and SDS concentration were all found to give increases in the micelle size, the aggregation number attaining values above 1500. Considerable efforts were devoted to the elucidation of the shape of the large micelles. The intensity of the scattered light varies with the hydrodynamic radius close to expectations for rod-like micelles. Recently, the angular dissymmetry of the scattered light was used to obtain the radius of gyration of the micelles[34]. As can be seen in Fig. 4.2, this quantity varies with the hydrodynamic radius as expected for rod-like micelles while other relations are expected for other geometries. A further quantity which could be obtained from the QLS results was the degree of micellar polydispersity. For the SDS micelles, a significant degree of polydispersity was noted; the degree of polydispersity is consistent with Mukerjee's work[200]. Rohde and Sackmann[202] used the QLS technique to study SDS solutions in the absence of added salt. The diffusion coefficient was found to increase with increasing SDS concentrations with a dip between 20 and 25 mM. The size of the micelles as deduced from the Stokes-Einstein relation was unreasonably small. These observations seem to correlate well with what can be predicted from the work of Berne and Pecora[175]

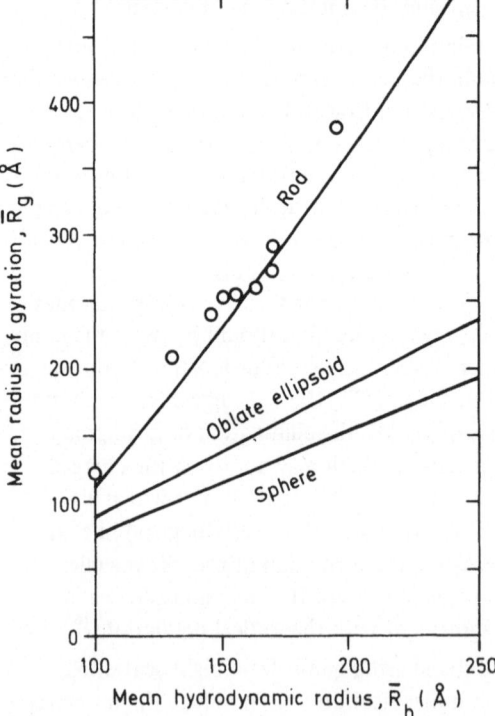

Fig. 4.2. Mean radius of gyration plotted versus the mean hydrodynamic radius for different micelle shapes according to theoretical predictions (*solid lines*). Circles give experimental quasi-elastic light scattering results for SDS micelles at different temperatures in 0.6 M NaCl. (From Ref.[34])

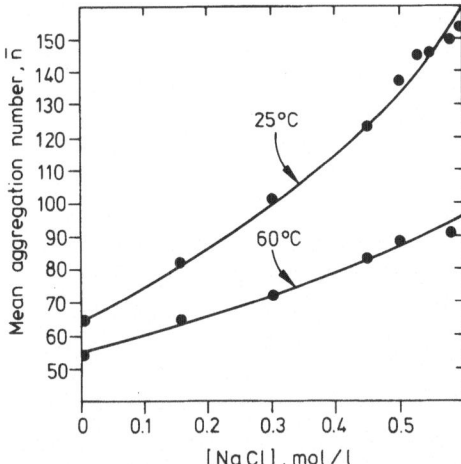

Fig. 4.3. The mean aggregation number of SDS micelles according to luminescence quenching studies plotted as a function of the NaCl concentration. The SDS concentration was 0.070 M. (From Ref.[203])

(cf. above) and suggest that the QLS method has considerable disadvantages in the absence of added salt.

Another interesting method for determining micelle aggregation numbers, in which luminescence quenching provides a "count" of the number of micelles, was recently applied to SDS[203]. While for the small micelles, there is agreement between different methods, the luminescence (Fig. 4.3) gives very much lower values than QLS for the large micelles formed at low temperature and high NaCl concentration. Sedimentation velocity[204] and classical light scattering[205] studies agree with the luminescence results on the other hand. One difficulty with the luminescence quenching method is that for larger micelles the excited state life-time becomes comparable to the time for diffusion over the micelle, which may lead to too low values of the micellar size.

4.3 Internal Structure of Micelles

The unterstanding of amphiphile association clearly must include detailed knowledge of the internal structure and dynamics, e.g., what is the conformation of the alkyl chains and what are their flexibility and packing conditions; is the interior of micelles exclusively of hydrocarbon nature or is there any water penetration? We will here consider the state of the hydrocarbon chains and defer a discussion of water penetration to the section on hydration.

Much of the early discussion of micelle structure centered around the problem of whether the hydrocarbon part should be considered as solid-like or liquid-like, the latter referring to conditions similar to those in liquid alkanes. Up to high alkyl chain lengths, the melting points of the alkanes lie below ambient temperature thus providing an indication that there is a liquid-like interior. However, the constraint offered by the micelle surface may, of course, substantially change the conditions.

47

Most useful and reliable information on these matters is obtained by considering the liquid crystalline state which is well established for higher surfactant concentrations. These liquid crystals are characterized by disorder over small distances and from calorimetric[206], X-ray[84] and NMR[207–209] measurements it is clear that there is a high degree of alkyl chain mobility; a direct evidence is that the sharp reflection at 4.3 Å characteristic of crystalline hydrocarbon chains is not detected in the X-ray studies of the liquid crystalline phases[84]. Recently, it was directly shown that the molecular motion in the rod-shaped micelles of CTAB at high concentrations is approximately the same as in the hexagonal liquid crystalline phase both for the amphiphile[194] and for solubilized benzene[67]. The same similarity between the molecular motion in micelles and in a hexagonal liquid crystalline phase has been observed for SDS using the ESR spin-label technique[210]. Concluding this discussion it may be said that for the surfactant liquid crystalline phases, a liquid-like state of the hydrocarbon chains has been well demonstrated and, furthermore, there is very strong evidence for a close similarity in this respect between the micellar solutions and the mesophases formed at higher surfactant concentrations. Since micelles and liquid crystals start to form at approximately the same temperature it is thus safe to conclude that the micelle interior has a liquid-like character. This idea was originally put forward in the pioneering work of Hartley from solubilization studies[20, 92]. He argued that the ability of micelles to solubilize large amounts of nonpolar substances can only be understood if the micelle interior has properties similar to those of liquid hydrocarbons; hydrocarbon chains packed as in a crystal do not have this ability. After Hartley's work other thermodynamic measurements[142, 211, 212] have also demonstrated a close similarity between the micelle core and liquid hydrocarbons.

Even if there is good agreement about the view that the micellar interior is 'liquid-like' rather than solid-like, there is confusion and conflicting ideas as regards the quantitative aspects. (To some extent, this is due to the use of unsuitable experimental approaches). Mukerjee[73, 213] inferred a partial solid-like character from odd-even alkyl chain length effects on the free energy of micelle formation; odd-even effects are important in the solid state. Most significant are the studies of partial molar volumes and compressibilities of amphiphile alkyl chains and of solubilized alkanes which demonstrate conditions very close to those in liquid alkanes[32, 40]. For example, the partial molar quantities of solubilized alkanes are virtually identical to those of the pure liquid alkanes. Direct measurements of the mobility of the micelle interior have been attempted by various spectroscopic methods; often one uses the ill-defined and unsuitable concept 'microviscosity'. Due to the strong solubilizing power of the micelles, it is a simple matter to introduce a spectroscopic probe. The fluorescence depolarization studies[214] of solubilizates like perylene and 2-methylanthracene have given high mobilities of these probes in micelles, although they are somewhat lower than those of the corresponding hydrocarbon solutions. Measurements of the ESR line widths of nitroxide spin labels solubilized in SDS micelles have given reorientational correlation times which are much shorter than for micelle rotation[215]. The spin label, therefore, appears to be free to reorient in the micelle but this reorientation is, according to the interpretation made, much slower than that of the free aqueous nitroxide spin label. In the present authors' opinion, significant criticism can be raised against ESR, fluorescence depolarization and similar methods. Firstly, the mobility of a foreign probe, that may appreciably perturb the system, is monitored and not that of the surfactant. In contrast to what has often been assumed, the probes used are probably located close to the micellar surface rather than in the micellar core; the molecular mobility in the head-group region certainly differs markedly from that in the center. Another aspect is that the interpretation is made in terms of a single rotational correlation time, whereas the process considered includes a partial averaging of the physical interaction by rapid local motions while the residual interaction is influenced by a slow motion over the dimensions of the micelle (see below). The distinction between the two correlation times has not been made.

An objection-free approach to these problems is to study the surfactant itself without using foreign probes. This can be achieved by using different NMR methods of which ^{13}C NMR so far has been most informative. ^{13}C chemical shifts of alkyl chains are dominated by conformation effects, i.e., they are mainly given by the trans-gauche ratio. For dilute aqueous solutions, the alkyl chain conformation is expected to correspond to a partly coiled state to reduce the hydrocarbon-water contact and for pure hydrocarbons, entropic effects, similarly, should favor coiled conformations. For a micelle, for geometrical reasons some chains have to be in the extended all trans conformation but the ^{13}C chemical shift studies indicate that the change in gauche-trans ratio on micelle formation is rather small. It is only of the order of 20% or less of the change from all gauche to all trans with the effects being largest in the center of the chains[50]. (An increasing trans-to-gauche ratio on micelle formation is also given by Raman studies[70], but there are also some conflicting results from the two techniques which remain to be resolved). ^{13}C NMR shifts of alkyl chains monitor very sensitively conformational effects as can be seen from recent studies[216] of mixed micelles of $CH_3(CH_2)_{13}\overset{+}{N}(CH_3)_3$ and $CH_3(CH_2)_{15}\overset{+}{N}(CH_3)_3$ salts. The ω-methyl signals of the two surfactants are shifted by a small amount relative to each other in the mixed micelle while they have the same shifts for the pure micelles. In the mixed micelle, the C_{14} chains can have a more extended conformation while the C_{16} chains are slightly compressed. Similar packing disturbances were found when decanol is solubilized in octanoate micelles[217].

^{13}C relaxation should be a most direct approach to the study of alkyl chain mobility in micelles and indeed a large number of studies have been presented. For hydrogen-binding carbons, the relaxation is dominated by intramolecular dipole-dipole interactions and on the basis of a standard expression, the correlation time has been deduced. The correlation time is a few times longer in the micelles than for the monomer in aqueous solution and there should thus be a marked slowing down of the alkyl chain motion on micelle formation. However, it was recently pointed out[218] that one cannot view the micellar interior as a simple isotropic medium. This is perhaps best illustrated by recalling that the micellar aggregate is structurally closely related to the liquid crystalline aggregates that are formed by amphiphilic substances at lower water contents. In these liquid crystalline phases, the molecular motion has been shown to be rapid as for example measured by diffusion[219]. However, the systems are usually anisotropic so that the average value of any tensorial quantity is nonzero, as illustrated, for example, by splittings in the deuterium NMR signal[67]. The local molecular packing is expected to be similar in a micelle so that the motion of the amphiphile or of a solubilizate is locally anisotropic. This introduces a complication in the analysis of experiments that are sensitive to the reorientational motion. In the simplest approximation, one has to consider both a fast local motion and a slower reorientation of the whole micelle. Such an analysis was performed for ^{13}C T_1 data in the sodium octanoate-water system[218]. It was found that substantial contributions to the T_1 values originate from the aggregate motion, particularly for the carbons closest to the polar headgroup. A direct demonstration that a slow motion, which can be micelle rotation or amphiphile lateral diffusion, is important has been provided by the observation of a dependence of T_1 of ^{13}C on the magnetic field strength. For the local motion, corresponding to chain flexibility and trans-gauche isomerization, there was no detectable change when going from an aqueous environment to a micellar one. It thus seems that the motion within a micelle is even more liquid-like than has been inferred previously from spectroscopic data.

In conclusion, it is evident then that the state of the alkyl chains in micelles is certainly quite close to that of liquid alkanes. The average conformation is only slightly more extended than for the monomer in aqueous solution and as regards the dynamics there seems to be no unambiguous demonstration of any change on micelle formation. As regards the lateral mobility, it is significant to note that the amphiphile translational diffusion in amphiphilic aggregates has been found to be of the same order of magnitude as in a corresponding liquid solution[219].

4.4 Counterion Binding

Aggregation of ionic amphiphiles is opposed by headgroup electrostatic repulsions which, however, are balanced to a large extent by adsorption of counterions at the micellar surface. While a unified view of electrostatic effects is given in Sect. 6, we here discuss counterion binding as revealed by different experimental techniques and in particular we discuss ion specificity. It is commonly stated that ca. 60–70% of the counterions are 'bound' to the micelles. Binding in this context should, however, not be taken to imply that there is a certain number of specific sites for the counterions; rather counterion binding is described in terms of a continuous radial distribution function corresponding to a high counterion concentration close to the micelle and continuously decaying with increasing distance from the micelle. There is thus certainly no unambiguous distinction between bound and free counterions. With such a state of affairs it is not surprising that the degree of counterion binding (β), defined as the ratio of counterions and amphiphile ions in a micelle, is no experimentally well-defined quantity; rather it is expected to vary with the experimental method used. There are quite a large number of techniques which can be used to study counterion binding; schematically, they may be subdivided into thermodynamic, transport and spectroscopic methods. Thermodynamic methods are exemplified by ion activity measurements which monitor the counterion concentration far away from the micelle. Transport methods monitor the amount of counterions moving with the micelle as a kinetic entity and are exemplified by tracer self-diffusion measurements. Spectroscopic methods monitor the amount of ions which have their spectroscopic properties affected by the micelles; examples are NMR chemical shift or relaxation studies. Examining the available information on counterion binding to micelles, some general conclusions can be made for simple hydrophilic counterions:

a) For a large number of systems, β lies in the range 0.5–0.8.

b) The three types of experimental approaches are in qualitative but not in quantitative agreement.

c) Variations in β with counterion are small for a given amphiphile.

d) Ion competition has been observed but is generally small.

e) Ion specificity may be marked as regards certain properties such as micelle size and shape but is small for the CMC.

f) β is only little dependent on factors such as surfactant concentration, temperature and alkyl chain length as long as the micelle shape remains the same.

g) At the transition from spherical to rod-like micelles, there is a small but significant increase in β and the same applies at transitions to liquid crystalline phases.

h) Solubilization in certain cases gives a small reduction in β, in other cases no effect is noticeable.

i) Ion association to micelles is badly accounted for in terms of conventional views of chemical equilibria but has many features in common with that of cylindrical polyelectrolytes. For example, the ion condensation concept has its counterparts for amphiphilic systems (see Sect. 6).

The somewhat unexpected fact that the different types of methods give approximately the same values of β is discussed in Sect. 6. Quantitatively, there are significant differences though, so in comparisons between different amphiphile ions, counterions etc., the adherence to a single method is necessary. Having outlined the generalities of counterion binding, we will with selected examples discuss some of its features on a molecular level as revealed mainly by spectroscopic studies.

Tracer diffusion studies have given β values close to 0.6 for sodium ion binding to a number of anionic amphiphile micelles, i.e., octanoate[220], dodecylsulfate[221], and octylbenzenesulfonate[222]. For CTAB, β is ca. 0.7[197] and both in this case, and in the case of sodium octanoate, a very slight increase in β with increasing concentration was noted. Membrane electrode studies by Vikingstad et al.[41] have, for sodium alkanoates, demonstrated a small increase in β with increasing alkyl chain length; β is found to be 0.57, 0.60, 0.64, 0.68, 0.72 and 0.74 for sodium heptanoate and upwards.

Mukerjee et al.[27] inferred a slightly lower β of Li^+ than of Na^+ for dodecylsulfate micelles and from this as well as from the variation of CMC with alkali counterion it was suggested that counterion binding follows the effective radius of the hydrated ion. For tetraalkylammonium ions, binding was found to increase with increasing ion size. For octylbenzenesulfonate micelles, competition experiments monitoring the Na^+ self-diffusion showed alkali ion binding to follow the sequence $Li^+ < Na^+ < K^+ < Rb^+ < Cs^+$ [222]. For alkanoates, the reverse sequence seems to apply as can be inferred from a slightly increasing CMC[223] along the series $Na^+ < K^+ < Rb^+ < Cs^+$. An increasing binding affinity with decreasing alkali ion atomic number for $-CO_2^-$ is consistent with a higher β for Na^+ than for K^+ for dodecanoate micelles[43] as well as with results from competition experiments using $^{23}Na^+$ quadrupole splittings of lamellar liquid crystals composed of alkali octanoate, decanol and water[224].

In NMR studies of alkali ion binding, a conspicious feature is the opposite $^{23}Na^+$ chemical shifts observed[225, 226] on micelle formation of sodium octanoate and sodium alkylsulfates (and sulfonates) (Fig. 2.17). Hydrocarbon chain length, on the other hand, does not appreciably affect either β or the intrinsic chemical shift of the bound ion for sodium alkylsulfates. Also several other NMR observations point to a marked headgroup specificity in alkali ion binding. Thus, both the $^{23}Na^+$ quadrupole relaxation in micellar solutions[226] and the quadrupole splittings in mesophases[63] show a markedly different behavior for $-CO_2^-$ and $-SO_4^-$ end-groups. The peculiar temperature and concentration dependences of the quadrupole splittings for the case of octanoate have been referred to an appreciable location of Na^+ ions between surfactant headgroups. For the case of $-OSO_3^-$ and SO_3^-, a location of Na^+ symmetrically with respect to the three oxygens pointing out into the aqueous layer fits the observations. The different locations in the two cases were found to be in conformity with the areas per polar group deduced from X-ray diffraction data. These observations suggest a model of counterion binding in these systems which applies also for other amphiphile aggregates since the interaction pattern is certainly independent of aggregate geometry[57]. Alkali ions appear in general to retain their primary hydration sheath on binding (see below) and it is proposed that hydrogen-bonding to the waters of alkali ion hydration is significant for the more basic $-CO_2^-$ group but not for the $-OSO_3^-$ or $-SO_3^-$ groups[226]. The simple picture, giving an enhanced counterion binding for a smaller radius of the hydrated ion, should thus be applicable for $-SO_4^-$ and $-SO_3^-$, while it should be modified for $-CO_2^-$. The suggested model is consistent with both the various NMR observations, with the location of counterions between $-CO_2^-$ end-groups and with the possibility of having opposite counterion binding sequences.

Counterion specificity has been observed to be markedly more pronounced for cationic surfactants than for anionic ones. This can certainly be mainly referred to a weaker hydration of typical counter-anions. From the variation of CMC with counterion and from ion activity measurements it can be inferred that the binding to $-\overset{+}{N}(CH_3)_3$ and $-\overset{+}{N}H_3$ headgroups follows the sequence $I^- > NO_3^- > Br^- > Cl^-$. (As an example a recent study[223] of decylammonium salts shows the CMC to decrease from 0.064 M for the chloride to 0.038 M for the iodide). The counterion specific effects on micellar shape for $-\overset{+}{N}(CH_3)_3$ surfactants were discussed above. For cationic (as well as some anionic) amphiphiles, a marked counterion specificity is also indicated in the phase diagrams[8] but systematic studies of the counterion dependence have not yet been reported.

Because of the possibility of charge-transfer interactions between polar head and halide ion, ion specific interactions can be expected to be particularly marked for alkylpyridinium halides. From the CMCs counterion dependence[3], as well as from counterion dissociation studies, binding is found to follow the sequence $I^- > Br^- > Cl^-$. The size of hexadecylpyridinium micelles is very sensitive to the anion of added salt, aggregation being promoted according to the sequence[227] $F^- < Cl^- \ll Br^- < NO_3^- \ll I^-$.

The behavior of hydrophobic counterions is, not unexpectedly, different from that of small hydrophilic counterions. Increasing counterion binding with increasing counterion size can be deduced both for anionic[27] and cationic[30] surfactants. Furthermore, the surfactant ion residence time in the micelle is markedly increased with hydrophobic counterions[30]. These and other observations point to a micelle-stabilizing effect due to hydrophobic interactions between surfactant and counterion.

4.5 Hydration

Micelle formation, being associated with eliminating the unfavorable contact between hydrophobic groups and water, is an event specific to aqueous solutions and an important matter refers to exactly how much of the amphiphile-water contact is retained on micelle formation. That the major part of the hydrocarbon chain-water contact is eliminated on micelle formation is an inherent aspect of our view of the driving forces of micelle formation but is also directly demonstrated by a large number of experimental techniques. Thus the water ^{1}H chemical shifts[228], the water ^{1}H relaxation rates[60] and the water self-diffusion coefficients[220] change at a relatively high rate below the CMC but more slowly above this concentration. Such variable concentration studies allow under certain assumptions the deduction of a global or effective hydration number of the micellized amphiphile but besides this we should also like to know, inter alia, the state of hydration of the counterions, the type of amphiphile headgroup-water interaction, as well as the dynamics of hydration water (binding life-time, rotational freedeom etc.).

The life-time of a monomer in a micelle may be of the order of a microsecond and in view of the accepted dynamic state of a micelle this implies that less extensive motions occur on a shorter time-scale. Aniansson[229] has recently examined the dynamic protrusion of methylene groups from the hydrocarbon core of a micelle

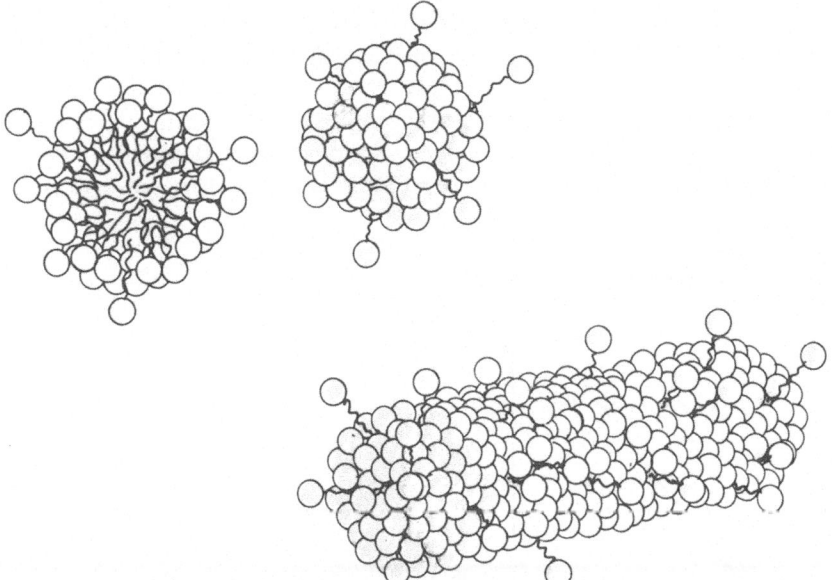

Fig. 4.4. A schematic picture of the dynamic protrusion of surfactant monomers from spherical and rod-shaped micelles. (By the courtesy of J. Ulmius)

on the basis of the kinetic information and the hydrophobic bonding energy. His results indicate a considerable protrusion. Thus every third monomer would protrude more than one methylene group and the average protrusion would be one $-CH_2-$ group. Even if it is open to some quantitative refinement, the work of Aniansson gives a good qualitative account of the extent of partial exits of monomers from micelles; in discussion of water penetration into micelles these results are most appropriate. An attempt to visualize the dynamic roughness of the micelle surface is given in Fig. 4.4.

The micelle-water interaction is highly dynamic and stoichiometrically undefined and, therefore, the concept of a single hydration number is a simplification; the meaning of the hydration number is dependent on the experimental approach considered. An initial suitable definition for our present purpose is to take the micelle hydration number as the number of water molecules moving with the micelle as a kinetic entity; this hydration number can be deduced from transport properties, e.g., viscosity and diffusion. A very useful description of the evaluation of hydration numbers from viscosity data has been given by Mukerjee[230]. The procedure involves the determination of the intrinsic viscosity and comparing it with the partial specific volume of the amphiphile; precision is reduced by certain corrections, mainly for electroviscous effects, which have to be applied. The micelle hydration numbers per amphiphile deduced by Mukerjee were 9 for SDS and 5 for both $C_{12}N(CH_3)_3Cl$ and $C_{14}N(CH_3)_3Cl$. A similar procedure was utilized to obtain a hydration number of 8.5–8.9 for C_7COONa micelles[231] and 10 for $C_{12}NO(CH_3)_2$ micelles[232]. Another study[233] obtained ca. 8 for SDS and for the series $C_{12}NO(CH_3)_2$ micelles[232]. Another study[233] obtained ca. 8 for SDS and for the series $C_{12}(OCH_2CH_2)_nOSO_3Na$ with n = 1–10 the hydration number was found to increase strongly with the number of ethylene oxide groups (Fig. 4.5). The study of the water self-diffusion coefficient as a function of micellar concentration is another efficient way of obtaining global hydration numbers and has given the hydration numbers 8.7 for C_7COONa[220] and 9 for $C_8(C_6H_4)SO_3Na$[234].

The hydration numbers given are somewhat approximative and are subject to error due to e.g., electroviscous effects and micelle shape effects. However, it seems that possible corrections should lower these numbers which can therefore serve as

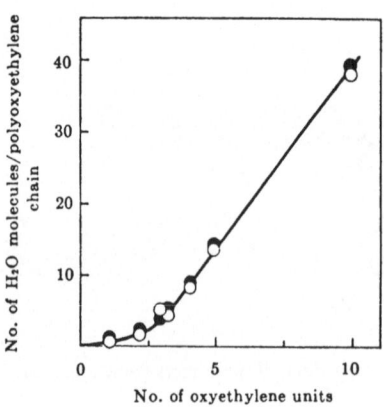

Fig. 4.5. The number of water molecules per polyoxyethylene chain as a function of the number of oxyethylene units for micelles of sodium dodecylpolyoxyethylene sulfates. (From Ref.[233])

rather reliable upper limits. The hydration numbers are smaller than those estimated for a uniform monolayer of water at the micellar surface and can be understood to a good approximation in terms of hydration of the bound counterions and the polar heads alone[4, 230]. Water contact of the alkyl chains is not indicated by these data. Very good evidence for the absence of any appreciable water penetration into micelles is provided by recent studies[32, 40] of partial molar volumes and partial molar compressibilities. For example, the partial molar quantities of alkanes in $C_{11}COONa$ solutions are close to the values of liquid alkanes but considerably higher than those in water. There exist suggestions in the literature of a considerable water penetration into micelles but these have been shown to be due to deficiencies in the procedures used as shown for example by Stigter[235] and by Mukerjee and Mysels[25]. Muller[236] has recently demonstrated that ^{19}F NMR chemical shifts, which have given the most important arguments in favor of a water penetration, are unreliable in this connection. Use of polar groups as spectroscopic probes (such as carbonyl in a recent ^{13}C NMR study[237]) are, of course, not likely to report faithfully on the unperturbed state of the micellar core. Furthermore, in many spectroscopic probe studies, the probe has often been erroneously assumed to lie in the micelle interior while in fact it resides close to the surface (see below).

Apart from the global hydration number, it is also of great interest to define the loci of hydration water and to study its mobility. Various spectroscopic methods, and in particular NMR, should be helpful in this respect but hitherto few adequate studies have been presented. A ^1H NMR study[238] of some nonionic polyoxyethylene-containing compounds showed that the alkyl chains are not in contact with water while the polyethoxy chains are strongly hydrated. A preceding ^1H relaxation study[239] of n-decylpentaoxyethylene-glycol monoether led to the same conclusion while ^1H chemical shifts were interpreted differently[240]. Recently, ^{19}F NMR relaxation studies[54] on $-CF_2-$ and $-CF_3$ groups in a number of both partially fluorinated and perfluorinated surfactants have been performed for D_2O and H_2O solutions. The difference in relaxation between the two solvents may be appreciable for the surfactant monomer, due to the much larger magnetic moment of ^1H, but disappears on micelle formation; this shows that the alkyl chain-water contact must be very small. Deuteron quadrupole splitting studies may provide information on the number of water molecules having their orientation affected by the amphiphile aggregates in liquid crystals. For the lamellar phase of the systems alkali octanoate-decanol-water, for example, at most about 5 water molecules per octanoate are appreciably oriented[64]. The low order parameter of the hydration water (ca. 0.01) indicates a high mobility, and, similarly, the water molecules at the surface of a micelle are certainly quite mobile. Tiddy et al.[241] have recently observed an ultrasonic relaxation for concentrated surfactant solutions and mesophases which they attribute to water exchange; these results are highly interesting and indicate water life-times at the surface of several different surfactant micelles to lie in the range $10^{-8} - 10^{-9}$ s.

Fundamental to our picture of counterion binding to micelles is a knowledge of whether counterions retain their hydration spheres or not. Mukerjee[242] concluded from partial molar volume data that (for simple ions) it is the interaction of the hydrated ion with the micelle that is important and the same conclusion is drawn in a

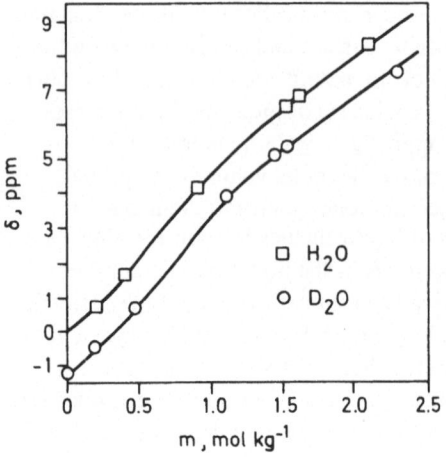

Fig. 4.6. ^{133}Cs chemical shifts of H_2O ($\square$) and D_2O ($\circ$) solutions of cesium octanoate as a function of amphiphile molality. A positive chemical shift is downfield. (From Ref.[59])

recent volumetric study[223]. An early ^{23}Na NMR relaxation study indicated that the Na^+ ion retains its inner hydration layer down to quite low water contents in surfactant systems[243]. A recent study[59] of the water isotope effect in Cs^+ shielding provides more direct evidence for an extensive hydration of alkali ions bound to micelles (Fig. 4.6); the observations of alkali ion dependent water deuteron quadrupole splittings for lamellar mesophases[64] as well as of alkali ion dependent phase diagrams[244] point in the same direction. As regards the halide ions, the observation of charge-transfer interactions for I^- [245–247] and of large changes in Cl^- and Br^- quadrupole relaxation rates[248] on binding to micelles are indicative of direct counterion-head-group approaches. The marked counterion specificity observed in certain cases (for CMC, micellar size and shape etc.) is, furthermore, difficult to unterstand if the counteranions completely retain their hydration layers. It is thus possible that at least Br^- and I^- ions (but perhaps also Cl^-) become partially dehydrated but there exists certainly no conclusive evidence. Thus Mukerjee et al.[73] and Stigter[249] consider that Cl^- ions retain their hydration water on binding to micelles and Kale and Zana[223] in their study of decylammonium micelles come to the same conclusion for both Cl^-, Br^-, I^- and NO_3^-.

Concluding our discussion on micelle hydration, we may state that the polar head-groups are certainly hydrated although to a varying extent and that alkali counterions retain their primary hydration sheath; small hydrophilic counteranions may at least in certain cases become partially dehydrated. There is no appreciable water penetration into the interior of micelles although the dynamic state of the micelle with frequent partial exits demands a limited water contact for the alkyl chains. This water contact certainly varies from system to system but may typically be around 50% or less for the α-CH_2 group and below 20% for the β-CH_2; it diminishes then rapidly as one moves away from the polar head. Evidence for a very low water content at some distance from the polar head was presented above and is, furthermore, provided by the slow passage of water through lipid bilayers[250] or between reversed micelles[251].

4.6 Solubilization

Solubilization is one of the most striking aspects of surfactant solutions. Here we will schematically consider the site of solubilization in a micelle although it may be more appropriate to consider the problem in terms of a mixed solubilizate-surfactant aggregate. If polar groups are present in the solubilizate a location close to the micellar surface is expected and for an amphiphilic solubilizate, such as a long-chain alcohol, surface location of the polar group and an orientation of the alkyl chain towards the micelle interior may be anticipated; for an alkane one may assume that location close to the micelle center is most favorable. These predictions are in accordance with, for example, the observation of a decreased $^{81}Br^-$ relaxation[58] on addition of hexanol, but not of cyclohexane, to concentrated CTAB solutions. Furthermore, n-alkanes solubilized in sodium dodecanoate solutions have partial molar volumes and compressibilities close to those of liquid alkanes[40].

For aromatic compounds it is much more difficult to predict the solubilization site. A large number of experimental investigations have related to this problem and there has been a considerable controversy about the dynamic solubilization site. However, there is now fairly general agreement that aromatic compounds such as benzene, naphthalene and pyrene are solubilized close to the amphiphile polar heads; alkyl substitution of the solubilizate moves the phenyl group somewhat away from the surface[73]. Aromatic compounds give sizeable 'ring current' shifts in NMR which were used by Eriksson and Gillberg[56] to study solubilization in CTAB solutions by ^1H NMR. Benzene, N,N-dimethylaniline and nitrobenzene were found at low solubilizate contents to lie close to the micelle surface while isopropylbenzene is drawn more inwards. Recently, 1-methylnaphthalene was found to lie close to the micelle surface for CTAB solutions; as can be seen in Fig. 4.7, the groups in the polar headgroup region are shifted much more than the others on solubilization[216]. The $^{81}Br^-$ NMR relaxation rate of CTAB solutions may be considerably reduced on solubilization of benzene and N,N-dimethylaniline. Data from other spectroscopic methods have been more difficult to interpret and various views have been advanced. However, recent studies by Mukerjee et al.[73, 252, 253] resolve the problems and show that benzene and some of its alkyl derivatives as well as naphthalene are located pre-

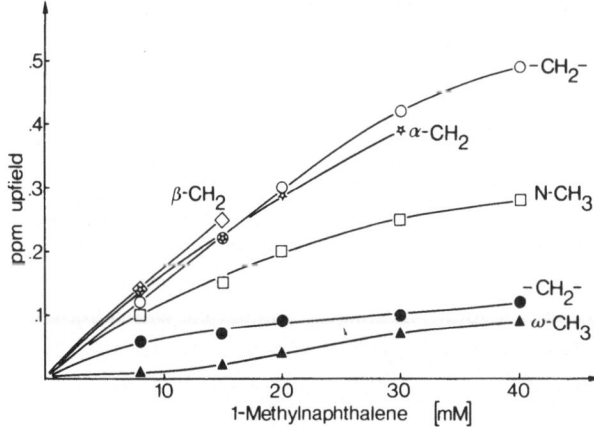

Fig. 4.7. NMR chemical shifts of CTAB protons as a function of 1-methylnaphthalene concentration. CTAB concentration 40 mM. (From Ref.[216])

dominantly in the interfacial region at low solubilizate contents. Solubilization in many cases induces very marked changes in micellar shape; it seems that such effects correlate well with solubilization in the head-group region[56, 58]. Solubilization of 1-methylnaphthalene in CTAB micelles, which occurs in this way[216], induces visco-elasticity and the same mode of solubilization is certainly operative for other cases of viscoelasticity[29, 198]. Solubilization of alcohols in CTAB solutions very drastically influences the rheological properties of the solutions; the effect varies in an irregular way with alcohol alkyl chain length[197].

We have already briefly touched on the relation between solubilization on one hand and counterion binding and hydration on the other. ^{81}Br NMR has shown a decreased counterion relaxation on solubilization in certain cases[58] but as shown by Br$^-$ self-diffusion data this may be referred mainly to an altered relaxation rate of bound Br$^-$ ions rather than a release of counterions[197]. For concentrated solutions of sodium octanoate, there is a gradual increase in β on solubilization of decanol[197] which correlates well with an increased ^{23}Na NMR relaxation[254]; specific complex formation between hydrated Na$^+$, $-$OH and $-$COO$^-$ at the micellar surface is a possible explanation. For SDS, there is no change in β on solubilization of cyclohexane, n-heptane and n-hexane while β is decreased on addition of benzene, p-xylene and octanol[197]; the effects are small, however. For sodium octylbenzenesulfonate solutions, solubilization of benzene and octanol leads first to a decrease in β and then an increase[255]. All the studies cited concern rather high solubilizate contents and it can be seen that the general pattern is that of a rather small influence of solubilization on counterion binding. The effect of solubilization on micelle hydration is much more sparsely documented. For approximately spherical CTAB micelles there is according to self-diffusion studies, a release of hydration water on solubilization of octanol and hexanol but not on solubilization of cyclohexane[256]; at 0.3 M CTAB, the water self-diffusion coefficient decreases on solubilization but in view of the concomitant shape change it is a little difficult to interpret these data. Solubilization studies using ^{2}H NMR have been performed for hexagonal mesophases of water and sodium octanoate or sodium octylsulfate[65]; a more polar solubilizate decreases water binding while a nonpolar one gives a small effect or an increase in water binding. One would in fact expect that a compound solubilized close to the micelle surface may displace some water while a compound solubilized in the micellar core may increase the surface per polar group and thus enhance hydration slightly.

5 Dynamic Processes in Micellar Systems

Intuitively it might be tempting to look upon a micelle as a rigid structure since one has a rather well-defined partitioning between the apolar interior and the aqueous environment. However, it was realized by Hartley[19, 92] at an early stage that the physico-chemical properties of micellar solutions are only compatible with a liquid-like micellar interior. Later investigations, using both thermodynamic and spectro-scopic techniques, have fully confirmed this conclusion. Today we can look upon the micellar aggregates as closely related to the liquid crystalline structures formed at higher amphiphile concentrations[8]. A consequence of the liquid-like structure of the micelle is that the equilibrium with monomers in the solution is a highly dynamical one. The amphiphiles constantly protrude from and leave the micelle[104, 229], while monomers in the solution recombine with the micelle. The same type of dynamic equilibrium also applies for solubilized molecules. For ionic amphiphiles, the counterions exchange rapidly between the micellar surface and the bulk solution. If there is a potentially chemically reactive species in the solution it can diffuse to the micelle where the reaction may be catalyzed. There is thus a number of dynamical processes going on in a micellar solution and studies of all these processes will add to our total picture of micellar systems.

5.1 Kinetics of Micelle Formation

During recent years the understanding of the kinetics of micelle formation has improved considerably. A thorough theoretical analysis of the problem was performed by Aniansson, Wall and co-workers[104, 187, 257−262]. The experimental methods were also developed to include not only ultrasonic relaxation[263−265] but also stopped flow[266, 267], p-jump[268−271], T-jump[272−275] and shock-tube[104, 276] methods. It was found that the theory of Aniansson and Wall was consistent with the experimental findings[104]. The crucial assumption in the theoretical model is that the elementary kinetic process that determines the dynamic behavior of the system is the equilibrium between a monomer and a micellar aggregate

$$A_1 + A_{s-1} \underset{k_s^-}{\overset{k_s^+}{\rightleftarrows}} A_s \qquad (5.1)$$

where k_s^+ and k_s^- are the 'on' and 'off' rate constants, respectively. A basic observation in the analysis is that micellization is a cooperative process. As a consequence the distribution curve for aggregate sizes has a strong minimum for aggregation numbers lower than the optimal micelle size. (See Fig. 2.23). The low equilibrium concentration of these aggregates of intermediate size provides one reason why one can neglect kinetic processes where aggregates A_p and A_r combine to the micelle A_{p+r}[277].

Consider a micellar solution at equilibrium that is subject to a sudden temperature change (T-jump). At the new temperature the equilibrium aggregate size distribution will be somewhat different and a redistribution of micellar sizes will occur. Aniansson and Wall now made the important observation that when scheme (5.1) represents the kinetic elementary step, and when there is a strong minimum in the micelle size distribution as in Fig. 2.23(a) the redistribution of micelle sizes is a two-step process. In the first and faster step relaxation occurs to a quasi-equilibrium state which is formed under the constraint that the total number of micelles remains constant. Thus the fast process involves reactions in scheme (5.1) for aggregates of sizes close to the maximum in the distribution. This process is characterized by an exponential relaxation with a time constant τ_1 equal to

$$\tau_1^{-1} = k^-/\sigma^2 + a\, k^-/n \qquad (5.2)$$

Here k^- is the 'off' rate constant for the micelle at the mean aggregation number n[262] and σ is the standard deviation of the assumed gaussian micelle size distribution. a is the relative amount of micellized monomer

$$a = (A_{tot} - \overline{A}_1)/\overline{A}_1 \tag{5.3}$$

where $\overline{A}_1$ is the equilibrium concentration of A_1. By measuring τ_1 as a function of a, i.e., of A_{tot}, one can determine k^-/σ^2 and k^-/n. If n is determined by a separate method, both k^- and σ can be calculated. Experimentally the value of τ_1 depends strongly on both surfactant concentration and the length of the alkyl chain but it is usually in the range $10^{-3}-10^{-8}$ s. This means that depending on the particular system different experimental methods must be used to determine τ_1. For the region $10^{-7}-10^{-8}$ s the ultrasonic absorption method is useful while in the range $10^{-3}-10^{-5}$ s p-jump techniques are applicable.

When the system has reached its quasi-equilibrium state a slower process, involving the relaxation to the true equilibrium, becomes measureable. This process involves a change in the number of micelles. The formation or dissolution of a micelle involves according to scheme (5.1) the appeerence of aggregates of size at the minimum of the size distribution curve, and since these aggregates occur with low probability the process can be a very slow one. Aniansson and Wall showed that this process is also characterized by an exponential decay with a relaxation time τ_2,

$$\tau_2^{-1} \simeq \frac{n^2}{\overline{A}_1} \frac{1}{R} \frac{1}{1 + \dfrac{\sigma^2 a}{n}} \tag{5.4}$$

Here R is a "resistance" for aggregate flow through the region of intermediate aggregate sizes

$$R = \sum_{s_1+1}^{s_2} \frac{1}{k_s^- \overline{A}_s} \tag{5.5}$$

where $\overline{A}_s$ is the mean concentration of aggregate A_s and s_1 and s_2 are aggregation numbers between the monomer and the proper micelles. In Eq. (5.4) it is R that is most sensitive to changes in external parameters such as temperature and salt concentration. Measurements of τ_2 thus give a unique insight into the properties of the aggregates close to the minimum in the distribution curve. Typically τ_2 decreases with increasing temperature and increases on the addition of salt[104]. This shows that the formation of the least probable aggregates is endothermic and that these aggregates are less stabilized by salt than the proper micelles. Typical values of τ_2 are in the range $10^{-3}-1$ s. There are thus several techniques that are applicable for determining the kinetics of the slow process.

The type of molecular information that can be obtained from measurements of the fast relaxation time is illustrated in Table 5.1 where data for a series of sodium alkylsulfates are presented. One can see that for micelles of short-chain surfactants,

Table 5.1. Rate constants and micelle size distribution for a series of sodium alkyl sulfates at 25 °C. (Adapted from Ref.[104])

Surfactant	CMC, M	n^a	σ^b	$k^- s^{-1}$	$k^+ M^{-1} s^{-1}$
NaC_6SO_4	0.42	17	6	$1.32 \cdot 10^9$	$3.2 \cdot 10^9$
NaC_7SO_4	0.22	22	10	$7.3 \cdot 10^8$	$3.3 \cdot 10^9$
NaC_8SO_4	0.13	27		$1.0 \cdot 10^8$	$7.7 \cdot 10^8$
NaC_9SO_4	6.10^{-2}	33		$1.4 \cdot 10^8$	$2.3 \cdot 10^9$
$NaC_{10}SO_4{}^c$	$3.3 \cdot 10^{-2}$	41		$9 \cdot 10^7$	$2.7 \cdot 10^9$
$NaC_{11}SO_4$	$1.6 \cdot 10^{-2}$	52		$4 \cdot 10^7$	$2.6 \cdot 10^9$
$NaC_{12}SO_4$	$8.2 \cdot 10^{-3}$	64	13	$1.0 \cdot 10^7$	$1.2 \cdot 10^9$
$NaC_{14}SO_4$	$2.05 \cdot 10^{-3}$	80	16.5	$9.6 \cdot 10^5$	$4.7 \cdot 10^8$
$NaC_{16}SO_4{}^d$	$4.5 \cdot 10^{-4}$	100	11	$6 \cdot 10^4$	$1.3 \cdot 10^8$

[a] Mean aggregation number;   [b] Standard deviation for the micelle size distribution; [c] 40 °C; [d] 30 °C.

the size distribution is rather broad ($\bar{n} \simeq 2 \sigma$), while for long-chain amphiphiles one has a relatively narrower distribution ($\sigma/\bar{n} \ll 1$). As a consequence of the broad distribution for the short chain amphiphiles, the long relaxation time is difficult to observe.

The rate constants k^- and k^+ in Table 5.1 are related through the CMC value and are thus not independent. In obtaining a molecular interpretation of the rate constants it seems most convenient to focus attention on the rate of the combination reaction, i.e., on k^+. In the absence of long-range forces, the maximum diffusion-controlled rate of a bimolecular reaction is

$$k_2 = 4 \pi N_a D r \qquad (5.6)$$

where k_2 is of dimension $M^{-1} s^{-1}$. N_a is Avogadro's number, $D(m^2 s^{-1})$ the sum of the diffusion coefficients of the reactants and $r(m)$ the distance from the micellar center at which the reaction occurs. From experimental values of the diffusion constant of the surfactant and estimates of micellar radii, the values obtained for k^+ using Eq. (5.6) are larger than the experimental ones by a factor of the order of ten; the discrepancy being largest for the amphiphiles with the longest chains. Since the micelle has a very dynamic structure it seems less likely that the micellar surface creates a barrier for the penetration of the monomer alkyl chain. A more plausible alternative, as suggested by Aniansson et al.[104], is that the monomers are repelled by the micelle through long-range electrostatic interactions.

It appears that, through a combination of theoretical and experimental efforts, the kinetic methods have developed into one of the most fruitful approaches to the study of physico-chemical properties of micellar systems. One of the main problems that remains to be solved is the kinetics for systems containing rod-shaped micelles where the size distribution is very broad. There is also a problem in the interpretation of the amplitudes measured in the different kinetic experiments.

5.2 Kinetics of Solubilizates and Counterions

One of the most important features of micellar solutions from a chemical point of view is their ability to solubilize otherwise water insoluble molecules. The liquid-like apolar micellar interior acts as a solvent for apolar substances. The solubilized molecules are of course also in dynamic equilibrium with the aqueous environment and other micelles. The kinetics of the solubilizate exchange has been studied by ESR methods using nitroxide radicals with a significant water solubility[278]. These studies indicated that the exchange process is rapid, but a detailed picture did not emerge. By the introduction of photochemical techniques, Thomas and co-workers[76, 78, 279] seem to have found an experimental method that makes it possible to study the various kinetic processes of solubilizates in detail. In micellar systems one has a spontaneous molecular organization that can be used advantageously to probe specifically different molecular processes. Take as an example the studies of the phosphorescence of solubilized 1-bromonaphthalene. After an excitation of the bromonaphthalene by a laser pulse, the resulting triplet state can relax to the ground state either through a radiative process or through an interaction with a triplet quencher. By having an ionic quencher with the same charge as the amphiphile forming the micelles one can ensure that the quenching process only occurs in the aqueous environment. By measuring the phosphorescence yield at varying concentrations of quencher and micelles it was possible[279] to determine the rate constant for exit and reentry of the 1-bromonaphthalene in sodium dodecylsulfate micelles.

The rate constant for the reentry is of the magnitude expected for a diffusion-controlled reaction as in Eq. (5.6). This means that the exit rate is determined by the partition coefficient of the solubilizate in its triplet state between the micelle and the aqueous solution. Table 5.2 shows the exit rate constants k_1 for several systems. The water solubilities of the probes are also given to show the correlation between k_1 and the solubility in water. These studies give further support to the view that the micelle has a very dynamic structure, which makes it easy for the solubilizate to enter and leave the aggregate.

The exchange between bound and free counterions is usually very rapid which probably explains why there exist so few experimental determinations of this rate constant. For example, in NMR studies of counterions one invariably observes a rapid exchange between free and bound ions. However, in solutions containing cetyl-

Table 5.2. Exit rate constants k_1 for solubilizates in micellar systems. (From Ref.[77])

Compound	Surfactant	k_1 (s^{-1})	Solubility in water (M)
Anthracene	CTAB	$2 \cdot 10^2$	$4.6 \cdot 10^{-6}$
1-Bromonaphthalene	SDS	$2.5 \cdot 10^4$	
Naphthalene	SDS	$> 5 \cdot 10^4$	$2.6 \cdot 10^{-4}$
Biphenyl	SDS	$1.2 \cdot 10^5$	$4.9 \cdot 10^{-5}$
Benzene	SDS	$\geqslant 10^6$	$2.1 \cdot 10^{-2}$
CH_2I_2	SDS	10^7	$4.7 \cdot 10^{-3}$

pyridinium iodide micelles Grünhagen[280] observed a relaxation process with a rate constant of $1.5 \cdot 10^7 \text{ s}^{-1}$ (40 °C), which was interpreted as due to a counterion dissociation process. In this case there is, in addition to the electrostatic ion-ion attraction, a charge transfer interaction between the iodide ion and the pyridinium ring giving a stronger counterion binding. In the absence of such specific effects the rate constant is expected to be substantially larger. The motion of the counterions can then próbably be described by a diffusion equation where the effect of the long range electrostatic interaction has been added. In the presence of an electrostatic potential Φ the diffusion equation for an ion k with charge z is

$$\frac{\partial C_k}{\partial t} = - D_k \left\{ \vec{\nabla} \left(\frac{ze}{kT} C_k \vec{\nabla} \Phi \right) - \nabla^2 C_k \right\}. \tag{5.7}$$

For counterions, Eq. (5.7) predicts a larger rate constant for the association process than Eq. (5.6) since the ions are attracted by the micelle. An estimate using Eq. (5.7) and a solution of the Poisson-Boltzmann equation (cf. Sect. 6) gives $k_2 \simeq 10^{10} \text{ M}^{-1}\text{s}^{-1}$ for the counterion association.

5.3 Micellar Catalysis

It often happens that the rate of a chemical reaction is substantially enhanced in a micellar solution relative to that of a corresponding pure aqueous system. This phenomenon is called micellar catalysis and has received a considerable interest in recent years[7,281]. There are basically three different mechanisms that can explain the enhanced rates[282]. A trivial effect is that in the micellar system one can increase the concentration of a semipolar reactant giving an increased rate without affecting the rate constant: an effect that can be of practical significance. The second possibility is that, by performing the reaction near the interface between the aqueous environment and the apolar micellar interior, one might stabilize the transition state of the reaction increasing the rate constant. This seems to occur to a lesser extent than might be expected. For example, the rate constants for unimolecular reactions usually change by less than a factor of ten. The apparently most important cases of micellar catalysis occur for bi- and trimolecular reactions. By introducing micelles into an aqueous system one makes it microscopically inhomogeneous, so that the local concentrations can deviate substantially from the bulk ones. For a bimolecular reaction the rate will be determined by the product of the local concentrations. Depending on reactants and system this product can be either greater or smaller than that of the bulk concentrations.

Take as an example the experimentally most studied reaction in micellar catalysis, ester hydrolysis in a basic medium

$$\underset{\text{R–C–O–R'} + \text{OH}^-}{\overset{\text{O}}{\underset{\|}{}}} \rightleftarrows \underset{\text{R–CO}^- + \text{R'OH}}{\overset{\text{O}}{\underset{\|}{}}}$$

If the radicals R and/or R' are hydrophobic, the ester will be solubilized by the micelles and the hydrolysis will occur at the surface of the micelle. When the micelles are formed by a cationic surfactant, the local concentration of OH⁻ ions close to the micellar surface will be substantially larger than the bulk concentration. Thus, by using a micellar system, one has been able to assemble the reactants in some spatial region which makes the reaction a more probable event. In enzymatic catalysis this mechanism has been termed the proximity effect[283].

The catalytic effect for reactions involving an ionic reactant usually shows a strong dependence on the total amphiphile concentration. The maximal effective rate constant is attained at concentrations just over the CMC. Romsted[284] showed that this occurs due to the competition between the ion binding of the reactive ions (OH⁻ in the example above) and the counterions of the amphiphile. Recently, Diekman and Frahm[285, 286] showed that it is possible to rationalize the kinetic data by describing the ion distribution through a solution of the Poisson-Boltzman equation. (See Fig. 5.1).

One of the reasons for the present interest in micellar catalysis is the analogy with enzymatic catalysis. For example, with respect to the proximity effect discussed above, such an analogy seems fruitful, but it should be kept in mind that there

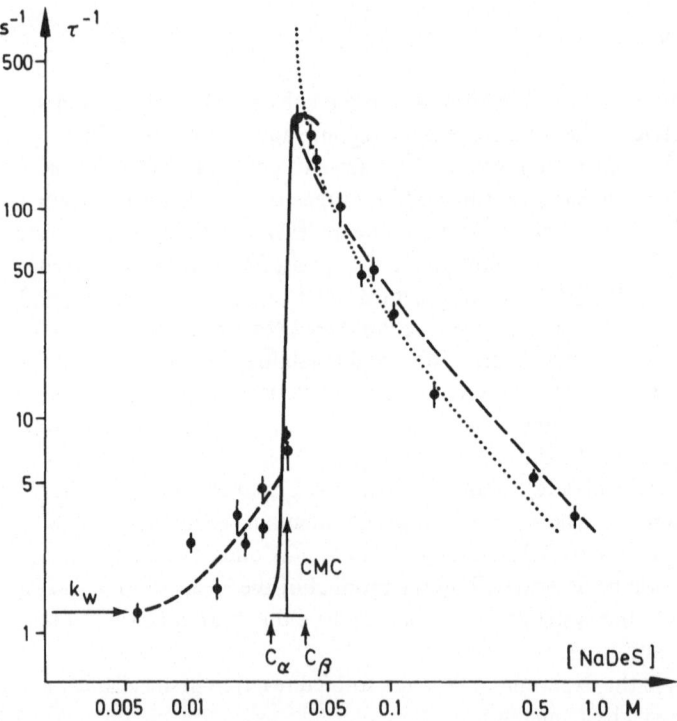

Fig. 5.1. Reaction rate for the complexation of Ni^{2+} and PADA (Pyridine-2-azo-p-dimethyl-aniline) in sodium decylsulfate (NaDeS) solutions. --- calculated rate from the Poisson-Boltzmann equation. · · · calculated rate using the assumption that all Ni^{2+} ions are bound to the micellar surface. (From Ref.[285])

are important differences between the two types of processes. In the micellar system one (or more) of the reactants are solubilized and the solubilization is a nonspecific dissolution process. In an enzyme, on the other hand, the substrate(s) is (are) bound to a specific site and this provides a basis for a much more efficient and specific catalytic process. It also follows that the kinetic equations are different in the two cases so that the reaction rate has a different functional dependence on the concentrations of reactants, a point that is neglected in simpler kinetic schemes[7] for micellar catalytic processes.

5.4 Translational Motion of Micelles

One aspect of the dynamics of micellar systems that has received a renewed interest during recent years is the translational motion of the micelles themselves. In the simplest approximation, the translational diffusion coefficient, D, of a spherical micelle is related to the hydrodynamic radius r_M through the Stokes-Einstein relation

$$D = kT/(6 \pi \eta r_M) \tag{5.8}$$

where η is the viscosity of the solvent. Thus by measuring D one should be able to determine the size of the micelle, a quantity that is surprisingly difficult to determine experimentally. In the limit where Eq. (5.8) applies mutual and self diffusion are equal.

For diffusion over macroscopic distances, D can conveniently be determined using radioactive or dye labelling[111, 287]. With the NMR pulsed gradient spin echo method[288, 289] one can determine the diffusion over distances of the order of $0.1-1 \mu m$. A third possibility is to analyze the quasi-elastically scattered light which is sensitive to fluctuations in the micellar position over very short times[33, 175]. If the system behaved ideally the three techniques would give the same value of D but in real solutions the micelle-micelle interactions are strong and these will have different effects in the different types of experiments.

For ionic amphiphiles in particular the micelle-micelle interaction is strong, but also nonionic micellar systems show strong deviations from ideality as the cloud-point is approached. Both repulsive and attractive forces will reduce the micellar diffusion coefficient as measured by the tracer and NMR techniques. That interactions influence D is seen experimentally from the concentration dependence of $D^{37, 289}$ (Fig. 2.11). It is also clear that Eq. (5.8) yields far too large values of r_M. The electrostatic corrections to D have been discussed by Mazo[290]. In the quasi-elastic light-scattering case the situation is more complex. The light scattering is caused by fluctuations on a short time-scale. For ionic amphiphiles it appears that, at least for low amphiphile concentrations, the process that dominates the scattering is the fluctuation of the micellar aggregate relative to the surrounding ionic atmosphere[291]. This process is a relatively fast one and the diffusion coefficient determined from the correlation time of the fluctuations in the scattering can be too large[202] giving too small radii, r_M. To avoid this effect, Mazer et al.[33] made measurements on sodium dodecyl sulphate in high salt concentration to contract the counterion charge cloud.

6 Electrostatic Effects

6.1 Introduction

An important class of micelle-forming amphiphiles is the ionic ones. In the monomeric state these ionic amphiphiles behave essentially as simple electrolytes and are
thus dissociated into the constituent ions in aqueous solution. As the CMC concentration is reached and micelles are formed the amphiphilic ions give the aggregate
a high formal charge. The high repulsion energy inherent in the aggregation of ions
of like charge is partly compensated by the binding of counterions. Nevertheless such
micelles are charged entities which is, for example, shown by their electrophoretic
mobilities.

The charged micelles give rise to strong local electrical fields in the solution,
which will influence the distribution and motion of other ionic entities. The micelles
also repel each other reducing the translational mobility. These electrostatic interactions influence the energetics of the micellization process substantially. This is
seen from the comparatively high values of the CMC for ionic amphiphiles and by
the fact that the addition of salt decreases the CMC.

It is thus clear that a treatment of the micellization process of ionic amphiphiles
must include a discussion of electrostatic effects. Furthermore, even for zwitterionic
and nonionic surfactants, the electrostatic effects play a role. The favorable interaction between the polar groups of these amphiphiles and the solvent water is probably mainly of an electrostatic origin.

6.2 Models for the Description of the Electrostatic Interactions

A system of an ionic amphiphile contains in its simplest form three entities: the solvent water, an amphiphilic ion, and a hydrophilic counterion. The properties of the
total system can then be understood as the effects of the mutual interactions between
these three species. A comprehensive treatment of all the interactions on a molecular
level is not at present feasible. However, the development of methods for determining intermolecular potentials and for making statistical mechanical simulations[292-294]
should change this in the not too distant future.

The most straightforward approach to the micelle formation is through equilibrium constants. For an ionic amphiphile the association can be described through a
number of equilibria

$$mM + nA \rightleftarrows M_m A_n \tag{6.1}$$

where M denotes the counterions. Here there is a large number of conceivable pairs
of values {m, n} to describe different aggregates. A description of the aggregation
with the use of (6.1) is designed for the discussion of changes in concentrations of
M and A. When describing a process in terms of a chemical equilibrium the entities
involved should be well-defined. This in turn presupposes short-range forces. For
simple electrolyte solutions, one corrects for the effects of the long-range electro-

static forces through activity coefficients. In polyelectrolyte systems, to which micellar solutions should be assigned, the long-range effects seem to dominate the behavior and to correct for these through activity coefficients seems illogical. A further problem with the description in scheme (6.1) is that when the counterion association is investigated experimentally different techniques will give different results since the exact nature of the aggregate $M_m A_n$ is unspecified.

In the phase separation model of micelle formation (cf. Sect. 3.1) it is also possible to include the counterions specifically. One has made the distinction between the uncharged phase and the charged micellar pseudophase[295]. These models can, for example, be used to predict how the CMC varies with salt concentration[46], but as used they are open to the same kind of criticism as is the equilibrium model.

Since the ion association in micellar systems is due to the long-range electrostatic interactions it is preferable to describe the ion distribution around the charged micelle explicitly. This is customarily made in a model where the water is approximated by a dielectric continuum in which the counterions are distributed. Within such a model, there is a multitude of different degrees of refinements. We choose here to describe the, in our view, most straightforward scheme and postpone a discussion of the possible modifications to a later stage.

Assume that the micellar aggregate is spherical (radius r_m) with the ionic groups of the amphiphile at the surface. Due to the dynamic nature of the micelle the

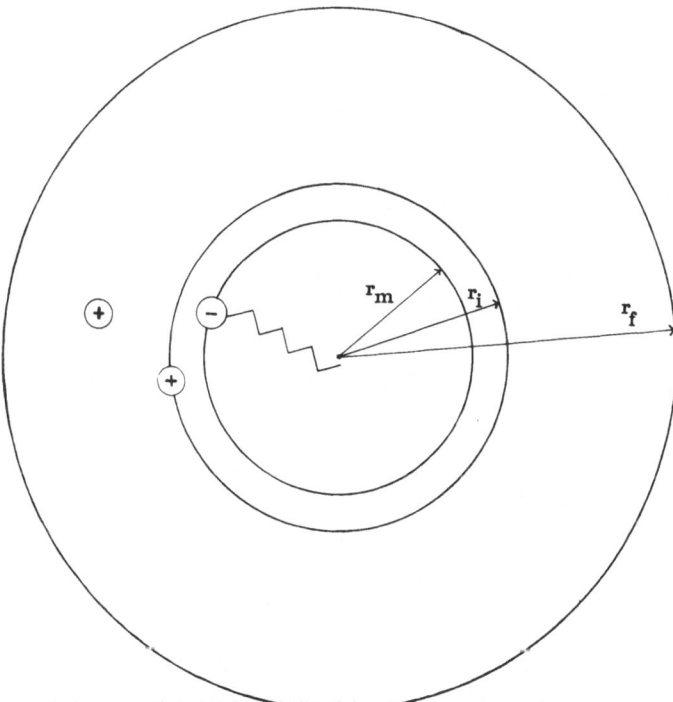

Fig. 6.1. Illustration of the model for the micellar system used in the calculation of the electrostatic properties. r_m is the radius of the micelle, r_i the radius for the closest approach of the counterions and r_f the radius of a sphere representing the volume per micelle

charges will on average be evenly distributed over the surface. In the surrounding medium, the counterions likewise acquire on average a spherically symmetrical distribution. Due to the finite size of the ionic groups and the counterions the center of the ions cannot approach within a sphere of radius r_i (see Fig. 6.1).

The ionic groups on the micellar surface and the counterions will give rise to a nonuniform electrostatic potential Φ according to the Poisson equation. If furthermore the electrostatic effects dominate the counterion distribution the ion concentration is determined by Φ following a Boltzmann distribution. These approximations lead to the Poisson-Boltzmann equation.

$$-\epsilon_r \epsilon_0 \, \nabla^2 \, \Phi = \rho = \sum_i c_i z_i e \, \exp(-z_i e \, \Phi/kT) \tag{6.2}$$

where ϵ_r is the relative permittivity, ϵ_0 the permittivity of a vacuum, ρ the charge distribution, c_i a constant with the dimension of a concentration and related to the mean concentration of ion i, z_i the valency of ion i and e the elementary charge. When one considers a single micellar aggregate in an electrolyte medium the constants c_i are the actual concentrations of the ions i if $\Phi(r \rightarrow \infty)$ is chosen to be zero. For this case one has the boundary condition

$$\frac{d\Phi}{dr} \Big|_{r \rightarrow \infty} = 0 \tag{6.3}$$

The second boundary condition is obtained from the micellar charge Q_m. It follows from Gauss' theorem that

$$4 \pi r_m^2 \, \epsilon_r' \epsilon_0 \, \frac{d\Phi}{dr} \Big|_{r = r_m} = 4 \pi r_i^2 \, \epsilon_r \epsilon_0 \, \frac{d\Phi}{dr} \Big|_{r = r_i} = -Q_m \tag{6.4}$$

where ϵ_r' is the dielectric constant at $r = r_m$. One can note that Eq. (6.4) is valid irrespective of the dielectric properties of the micellar interior. In the region $r > r_i$ Eq. (6.2) has to be integrated numerically. This was performed first by Hoskin[296] with the boundary condition of Eq. (6.4) replaced by a value for the surface potential. Extensive listings of solutions were made by Loeb et al.[297]. When applying these solutions to micellar problems, one should note that it is the charge of the micelle that is the constitutive property and that the surface potential $\{\Phi(r_i)$ or $\Phi(r_m)\}$ may vary with external conditions. The solution of Eq. (6.2) defines the potential $\Phi(r_i)$ and also the ion distribution, within the given approximation.

The electrostatic effects are influenced by the micelle concentration. This effect can be viewed as a micelle-micelle interaction mediated by counterions. The most direct way for modelling the finite micelle concentration is to confine the volume per micelle by an outer radius r_f of finite size[298–300]. This is called the cell model.

There are many possible improvements to the Poisson-Boltzmann equation and an extensive discussion of the refinements has been presented by Bell and Levine[301]. The relative permittivity is field dependent and the ions are polarizable. In Eq. (6.2) the correlation between the ions is neglected; so are specific chemical effects in the interactions between the ions and between the ions and the water. Furthermore the micellar surface is not fixed but the amphiphiles protrude[229]. In spite of all these shortcomings, it appears that Eq. (6.2) provides a good qualitative description of the electrostatic effects and that even many of the quantitative predictions are surprisingly accurate. The basic reason for this seems to be that the Poisson-Boltzmann equation correctly describes the long-range correlations.

6.3 Electrostatic Energies

From the solution of the Poisson-Boltzmann equation one can calculate the electrostatic contribution to the free energy. It is illustrative to divide G into two parts[302]. The first is concerned with the free energy of the electric field and is given by:

$$G_f = \frac{1}{2} \int_V \rho \Phi \, dV = \frac{\epsilon_r \epsilon_0}{2} \int_V (\vec{\nabla} \Phi)^2 \, dV \qquad (6.5)$$

where the integral is over the whole volume of the micellar system and ρ is the total charge density. The second equality in Eq. (6.5) presupposes electroneutrality in the volume V. The second contribution is due to the entropy of the counterion distribution which is affected by the electrostatic interactions:

$$S_e = -R \sum_i \left\{ \int_V C_i (\ln C_i - 1) \, dV - n_i (\ln \overline{C}_i - 1) \right\} \qquad (6.6)$$

where n_i is the total amount of ion i in the volume V, C_i the local concentration and $\overline{C}_i$ the mean concentration. In Eqs. (6.5) and (6.6) the energy and entropy have been given relative to the corresponding neutral system. For this case the total free energy $G_e = G_f - TS_e$ will always be positive. To actually calculate G_e one can proceed in several ways. Either the integrals in Eqs. (6.5) and (6.6) are calculated directly or one can determine the work that is required to charge the system to its final value

$$G_e = \int_0^1 2 \, G_f'(\lambda) \, \frac{d\lambda}{\lambda} \qquad (6.7)$$

where λ is the charging parameter and G_f' is given by Eq. (6.5). This method was used in the pioneering work of Overbeek and Stigter[303]. It is also possible to integrate a Gibbs-Helmholtz equation (ϵ_r constant)

$$G_e = T \int_0^{1/T} G_f' \, d(1/T). \qquad (6.8)$$

It was elegantly shown by Marcus[302] that within the approximations of Eq. (6.2) the three procedures are mathematically equivalent.

In a micelle formation process the volume changes are small and the equilibria are only slightly affected by the external pressure[41]. This means that the internal

69

energy and the enthalpy as well as the Helmholtz' and Gibbs' free energies are nearly equal, and their respective differences are neglected in the following. The free energy G_e can be decomposed into its enthalpic and entropic part using the Gibbs-Helmholtz equation and[304]

$$H_e = \frac{\epsilon_0}{2} \left(\epsilon_r + T \frac{\partial \epsilon_r}{\partial T} \right) \int_V (\vec{\nabla} \Phi)^2 \, dV \qquad (6.9)$$

For water the term $T \frac{\partial \epsilon_r}{\partial T}$ ($\simeq -105$ at 298 K) is large and negative[305] due to the strong orientation of the water dipoles in an electric field. This special property of water as a dielectric medium has two important consequences. Firstly G_f of Eq. (6.5) is mainly of entropical origin due to the cancellation between ϵ_r and $T \frac{\partial \epsilon_r}{\partial T}$ in Eq. (6.9). Since the contribution in Eq. (6.6) is purely entropical (and large) it follows the G_e is entropy dominated. Secondly H_e has a different sign than one would expect intuitively. Since at 298 K, $\epsilon_r + T \frac{\partial \epsilon_r}{\partial T} \simeq -26$ the enthalpy, according to the model, decreases when a charge is created on the micelle. The larger the micellar charge the larger are the electrostatic effects and the more negative is H_e on addition of a monomer to the micelle. This effect might be one of the explanations of the variation of $\Delta H_i'$ of Eq. (3.14) with aggregation number discussed in Sect. 3.5.

In applications concerning equilibria in the micellar solution it is often preferable to calculate the chemical potentials μ of the components instead of explicitly calculating the electrostatic free energy G_e. The point of using μ_i:s is that these are determined by the calculated ion concentrations at the border of the free micellar volume where the electrostatic potential and field are zero[302]. For an ion

$$\mu_i^{aq} = \mu_i^{\theta,\, aq} + kT \ln C_i \, (\Phi = 0) \, . \qquad (6.10)$$

For the water

$$\mu_{H_2O} = \mu_{H_2O}^{\theta} - kT \sum_i C_i (\Phi = 0) \qquad (6.11)$$

where the sum is over all ionic species. The amphiphilic ion can also reside in a micellar aggregate where its chemical potential is determined by the short-range molecular interactions and an electrostatic interaction, C_m micelle concentration[305]

$$\mu_A^{mic} = \mu_A^{\theta,mic} + G_e/n + (kT/n) \ln C_m \, . \qquad (6.12)$$

The chemical potential should be equal for the amphiphilic ion throughout the solution so that $\mu_A^{mic} = \mu_A^{aq}$, or

$$kT \ln C_A = (\mu_A^{\theta,\, mic} - \mu_A^{\theta,\, aq}) + G_e/n + kT/n \ln C_m \, . \qquad (6.13)$$

which is a relation between the monomer concentration of the amphiphile, the difference $(\mu_A^{\theta, mic} - \mu_A^{\theta, aq})$ determined mainly by the hydrophobic interaction and finally by G_e the electrostatic free energy. G_e is determined by the micellar radius and charge, by r_i and by the salt concentration. Thus G_e must be determined from the solution of Eq. (6.2). A shortcoming of Eqs. (6.10), (6.11), and (6.13) which is inherent in Eq. (6.2) is that the ion activities should contain a Debye-Hückel type correction but this might be added a posteriori. Eq. (6.13) provides a convenient starting point for a discussion of the variation of the CMC with various parameters, a problem that will be discussed further in Sect. 6.5.

6.4 The Ion Distribution

The micellar surface has a high charge density and the stability of the aggregate is heavily dependent on the binding of counterions to the surface. From the solution of the Poisson-Boltzmann equation one finds that a large fraction (0.4–0.7) of the counterions is in the nearest vicinity of the micellar surface[300]. These ions could be associated with the Stern layer, but it seems simpler not to make a distinction between the ions of the Stern layer and those more diffusely bound. They are all part of the counterions and their distribution is primarily determined by electrostatic effects.

In the theory of counterion binding in solutions of rodshaped polyelectrolyte ions, the concept of ion condensation[306, 307] has received much attention during recent years. The characteristic feature of the ion condensation model is that above a certain line charge density on the polyion the counterions 'condense' on the polyion to reduce the effective charge to a given value. It has been stated that the ion condensation behavior is peculiar for systems of rodshaped polyions, but it was recently shown that essentially the same ion binding properties apply also for planar[308, 309] and spherical systems[300, 310] under typical experimental conditions. One of the more intuitively unexpected manifestations of the ion condensation behavior is that the counterion concentration close to the polyion is only slightly affected by the mean salt concentration and the polyelectrolyte concentration. It also follows, in accordance with the observation that the ion binding is entropy-driven, that the ion distribution is rather insensitive to temperature changes.

Experimentally the ion binding can be investigated through three different kinds of measurements. Spectroscopic properties of the ions, or of probes interacting with the ions, are usually determined by short range interactions. The properties of the ions are then appreciably different from those of an ordinary aqueous solution only for those ions that are in direct contact with the micellar surface. This behavior is typical for the NMR[57, 59] and ESR[69] measurements on the counterions. Similar short range effects are also present in fluorescence quenching[78], charge transfer electronic absorption[247] and also in micellar catalytic reactions involving ionic reactants[284, 285]. The amounts of ions close to the micelle can be rather easily determined from the solutions of Eq. (6.2). A second approach to the ion binding is through transport properties like conductivity[27] and ion diffusion[220] but also light scattering[311]. The interpretation of these quantities is more complex and requires in principle an analysis of the whole ion distribution. An approximation that has been applied to distinguish between free and bound ions with respect to transport is to consider an ion bound if it is attracted electrostatically by more than kT by the micelle[300]. The virtue of such a definition is that the quantity of bound ions can be easily obtained from the theoretical calculations. The third way of determining the degree of counterion binding is through a thermodynamic quantity. This can involve direct measurements of the counterion activity using ion specific electrodes[46, 312] or a determination of the osmotic coefficient[313]. Such measurements can be interpreted directly using Eqs. (6.10) and (6.11)[299]. One can also attempt to study the ion binding by observing the variation of the CMC with salt, but severe problems of interpretation might be associated with the procedure as will be clear from Sect. 6.5.

In spite of the fact that the different types of procedures for determining the degree of counterion binding, β, may seem very different, calculations, using parameters typical for micellar systems, show that β has values in the range 0.4–0.7. That the different methods give similar values is also found experimentally[9, 27]. Furthermore, the agreement in the experimental and theoretical values of β is gratifying.

6.5 Variation of the CMC with Salt Concentration, Alkyl Chain Length and Amphiphile Ionic Group

Electrostatic interactions influence strongly the dependence of the CMC on a number of parameters. It is well established that the CMC decreases on addition of a simple salt to an amphiphile solution. Qualitatively, this is easily understood from the observation that the salt reduces the electrostatic repulsions and thus promotes micelle formation. A quantitative analysis of the problem can be based on Eq. (6.13). In this equation it is the free chergy G_e that varies with the salt concentration. The value of G_e can be obtained by solving Eq. (6.2) for various salt contents and calculating G_f and S_e according to Eqs. (6.5) and (6.6). One can note that it is the change in S_e that gives the largest contribution to the change in the electrostatic free energy on addition of salt, while changes in the direct ion – ion interaction are less important. In the limit of an infinite micellar radius, i.e. a charged planar surface, the salt dependence of G_e is solely due to the entropy factor. A difficult question when applying Eq. (6.13) to the salt dependence of the CMC is if Debye-Hückel correction factors should be included in the monomer activity. When G_e is obtained from a solution of the Poisson-Boltzmann equation in which the correlations between the mobile ions are neglected, it might be that the use of Debye-Hückel activity factors give an unbalanced treatment. If the correlations between the mobile ions are not considered in the ionic atmosphere of the micelle they should not be included for the free ions in solution.

An example of a calculation of the salt concentration dependence of the CMC on the basis of Eq. (6.13) is shown in Fig. 6.2. A fair agreement between theory and experiment is observed using only one adjustable parameter which fixes the absolute value of the calculated CMC. The slope ($\simeq$ .73) of the approximately straight line is

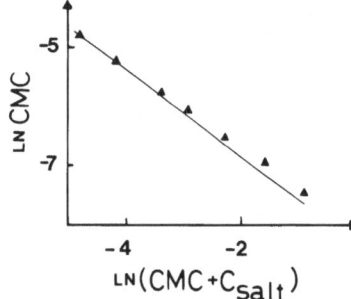

Fig. 6.2. A comparison between experimentally determined CMC values for sodium dodecyl sulfate (▲) (Ref.[186]) and the results of Eq. (6.13)

determined a priori through G_f and it is not in any direct way related to the degree of ion binding β as is sometimes suggested.

For nonionic surfactants there is, in a homologous series, a steady decrease in the CMC as the length of the hydrocarbon chain is increased. The decrease in the CMC corresponds to a gain in the hydrophobic interaction of about 1.1 kT per CH_2-group[4]. This value is only slightly smaller than the value 1.21 kT found from studies of solubilities of apolar substances in water. For ionic amphiphiles[4, 116], the CMC is much less dependent on the number of carbon atoms in the chain. The reason for this is found in the electrostatic effects. The monomeric amphiphile with its counterion acts as an electrolyte so that the higher the CMC the higher the effective salt concentration which as such decreases the CMC. The variation of the CMC in a homologous series of ionic amphiphiles in the presence of excess salt is very similar to that of nonionic surfactants[28, 114]. The dependence of the CMC on alkyl chain length can be discussed in quantitative terms using Eq. (6.13). As discussed above the G_e depends on the salt concentration, but it depends also on the radius and the surface charge density of the micelle. In the simplest model of a micelle the radius is equal to the length of the monomer in the all trans conformation. In that case, the area per ionic group is constant and thus also the mean surface charge density. By calculating the electrostatic free energy as a function of the salt concentration at constant micelle surface charge density for varying micellar radii one can estimate the dependence of the CMC on alkyl chain length.

One contribution, ΔG_e, to the electrostatic free energy comes from G_f in the region $r_m < r < r_i$. Since no charges are present in that region Eq. (6.5) can be integrated directly using Eq. (6.4) and the assumption of a constant dielectric permittivity to give

$$\Delta G_e = Q_m^2/(8\,\pi\,\epsilon_r'\,\epsilon_0)\,(1/r_m - 1/r_i) \quad . \tag{6.14}$$

Eqs. (6.14) and (6.13) in combination can be used to compare CMC values for amphiphiles with the same alkyl chain length but with different ionic groups. If it is assumed that the radius r_i and the charge of the micelle are constant in such a series, the variation in the CMC is due to the variation in r_m, i.e., how deeply the amphiphile charge is buried in the micelle. The smaller r_m, the larger ΔG_e, and the higher the CMC. This effect was thoroughly discussed by Stigter[235], using a somewhat different formalism, and it was shown that a reasonable rationalization of the variation of the CMC with ionic group can be obtained.

6.6 Electrostatic Aspects on Ion Binding Specificity

Micellar solutions of anionic amphiphiles are usually not stable with respect to the addition of di- or multivalent cations since a precipitation occurs (hard water). In exceptional cases, where precipitation does not occur, the question arises as to how the uni-, di- and multivalent ions compete for binding to the micelles. Due to the high value of $|\Phi|$ close to the micellar surface a counterion of high charge will be strongly favored and there is a discrimination between the different types of ions[299]. For example for SDS close to the CMC, $-e\Phi(r_i)/kT \simeq 7$ and if the presence of small amounts of calcium ions does not affect $\Phi(r_i)$, Eq. (6.2) can be used to calculate the ratio

$$\{C_{Ca}(r_i)/C_{Ca}(\infty)\}/\{C_{Na}(r_i)/C_{Na}(\infty)\} \sim 1100 .$$

An increase in salt and/or micelle concentration leads to a decrease in $|\Phi(r_i)|$ and the discrimination between univalent and divalent ions is decreased. The same effects should occur for cationic amphiphiles but they have been even less studied.

For ions of the same charge, the only parameter in the model presented above that can give rise to an ion specificity is the distance of closest approach $r_i - r_m$. In the model, the smaller ion gives the stronger ion binding. However, it seems that in most cases other factors, e.g., ion hydration and ion polarizability, dominate over the ion size effect, particularly for anions. (cf. Sect. 4).

6.7 Electrostatic Effects on Micelle Size and Shape

For ionic amphiphiles the first formed aggregates are closely spherical. At higher amphiphile concentrations there is a tendency for the formation of rod-shaped micelles[168]. Also the addition of salt favours the rod-shape aggregates[33]. It has been suggested that disc-shaped micelles also occur[160] but experimental evidence in favor of this view has only been obtained for mixed micelles of lecithin and sodium cholate[179].

Electrostatic ionic interactions can be seen to influence the preferred aggregate shape through two different mechanisms. Firstly, the electrostatic free energy is different for a sphere than for a rod at constant surface charge density. This difference varies with surfactant and salt concentration. Secondly, a change in aggregate shape also implies changes in the area per ionic group.

The analysis of the first effect can be made by comparing a spherical and an infinite rod-shaped aggregate with the same radius, surface charge density and amphiphile concentration. The infinite rod really corresponds to the condition in the nor-

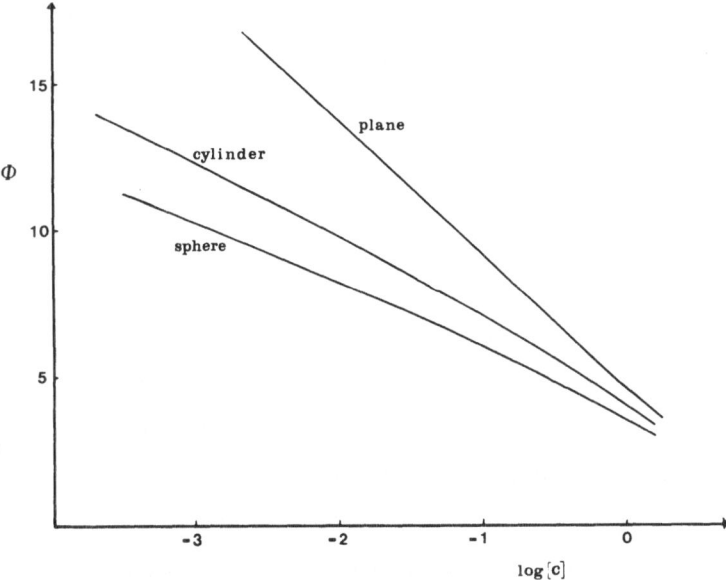

Fig. 6.3. The calculated reduced surface potential $\Phi = e\Phi(r_i)/kT$ versus the logarithm of the amphiphile concentration C (M) with no salt added for a spherical, cylindrical and planar aggregate. The surface charge density has been chosen as fixed at $\sigma_s = 0.228$ Cm^{-2}. The radii of the sphere and the cylinder are 1.8 nm

mal hexagonal phase[8] of the amphiphile system and the properties of a solution of finite rod-shaped aggregates should be intermediate between those of the spherical and the infinite cylindrical system. For the cylindrical system, Eq. (6.2) has an analytical solution in the absence of added salt[314]. A qualitative comparison between the two systems can be made under the conditions of no added salt. Figure 6.3 shows how the surface potential $\Phi(r_i)$ varies with surfactant concentration. The value of $\Phi(r_i)$ is always higher for the cylindrical aggregate but the difference decreases with increasing concentration. The necessary gain in the hydrophobic energy for the transition from a sphere to a rod to occur thus decreases as the amphiphile concentration is increased. This observation provides a partial explanation of why the tendency to form rod-shaped structures increases with increasing amphiphile concentration. The qualitative explanation of the effect is apparently that it is always more favorable electrostatically to assemble the amphiphiles in smaller aggregates with lower net charges. However, as the amphiphile concentration is increased, the interaggregate interactions become important reducing the difference between the two modes of aggregation. A second effect is illustrated in Fig. 6.4, which shows how $\Phi(r_i)$ varies with r_i for a given surface charge density and amphiphile concentration. As expected the difference between the sphere and the cylinder decreases as r_i increases. Thus amphiphiles with long alkyl chains are expected to have a larger tendency to form rod-shaped aggregates for simple electrostatic reasons. This is indeed confirmed by experiments.

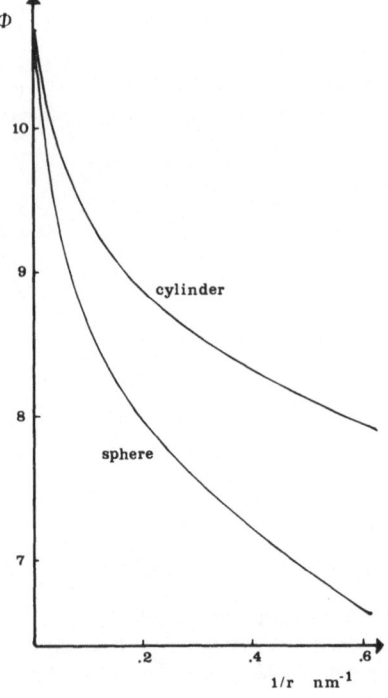

Fig. 6.4. The calculated reduced surface potential $\Phi = e\Phi(r_i)/kT$ versus the inverse radius for a sphere and a cylinder. Amphiphile concentration 45 mM and surface charge density $\sigma_s = 0.228$ Cm^{-2}

It is more difficult to discuss the difference between the areas per ionic group in a spherical and a cylindrical aggregate in quantitative terms. The alkyl chain volume per monomer is constant and if the radii of the aggregates remain the same this leads to a decrease in area per polar group when the cylinder is formed from spheres. The increase in charge density adds to an increase in the electrostatic free energy and this provides an additional electrostatic preference for the spherical system. When salt is added it has the general effect of diminishing the importance of the electrostatic contributions. Thus the electrostatic difference between the spheres and the rods is decreased, ultimately leading to the formation of rod-shaped micelles. For amphiphiles with long alkyl chains in particular such transitions are often found.

6.8 Micelle-Micelle Interactions

The micelles of ionic amphiphiles have high charges. A substantial part of this charge is neutralized by closely bound counterions but there are also less firmly bound ions which make the effective size of the micelle rather large. Micelle-micelle interactions can thus be rather long-range. As Fig. 6.3 demonstrates, micelle-micelle interactions have effects even at very high dilutions in the absence of added salt. In real systems, salt is always present and this leads to a screening of the interaction. One measure of the effect of the screening is provided by the value of the Debye screening length

$(1/\kappa)$. Far from the micelle, the electrostatic potential decreases as $\exp(-\kappa r)$, so that $1/\kappa$ is the length where the potential has decreased by $1/e$. However, one should remember that even such a screened potential can give rise to a substantial interaction. From calculations of how $\Phi(r_i)$ varies with micelle and salt concentration, it appears that the intermicelle interactions begin to become important when the concentration of counterions (from micellized amphiphile) is of the same magnitude as the salt concentration[300].

Effects of intermicellar interaction are seen experimentally in several ways. Of historical interest is that X-ray diffraction studies revealed long-range order in concentrated micellar systems, which was first interpreted[315, 316] as due to 'lamellar micelles'. Later, the long-range order was shown to be due to repulsion between the micelles[317]. A similar ordering effect probably causes the spectacular viscoelastic behavior of solutions where rod-shaped micelles are formed at low concentrations[172]. The electrostatic repulsions reduce the intensity of the light scattered by micellar solutions[200] and also influence the fluctuations in the scattering[291].

Acknowledgement. G. Gunnarsson is thanked for the calculations underlying Fig. 6.2.

7 References

1. Wennerström, H., Lindman, B.: Phys. Rep. *52*, 1 (1979)
2. Shinoda, K., et al.: Colloidal Surfactants. Some Physico-Chemical Properties. New York: Academic 1963
3. Mukerjee, P., Mysels, K. J.: Critical micelle concentrations of aqueous surfactant systems. NSRDS-NBS 36. Washington, D. S.: U. S. Governement Printing Office 1971
4. Tanford, C.: The hydrophobic effect. New York: Wiley 1973
5. Mittal, K. L. (ed.): Micellization, solubilization, and microemulsions. New York: Plenum 1977
6. Mukerjee, P.: In: Enriching topics from colloid and surface science, p. 135. Van Olphen, H., Mysels, K. J. (eds.). La Jolla, California: Theorex 1975
7. Fendler, J. H., Fendler, E. J.: Catalysis in micellar and macromolecular systems. New York: Academic 1975
8. Ekwall, P.: Adv. Liquid Cryst. *1*, 1 (1975)
9. Anacker, E. W.: In: Cationic surfactants. p. 203. Jungermann, E. (ed.) New York: Dekker 1970
10. Khetrapal, C. L., et al.: Lyotropic liquid crystals. Berlin, Heidelberg, New York: Springer-Verlag 1975
11. Nakagawa, T., Tokiwa, F.: Surf. and Colloid Sci. *9*, 69 (1976)
12. Kresheck, G. C.: In: Water, a comprehensive treatise. vol. 4, p. 95. Franks, F. (ed.). New York: Plenum Press 1975
13. Elworthy, P. H., Florence, A. T., Macfarlane, C. B.: Solubilization by surface-active agents. London: Chapman and Hall 1968
14. Durham, K. (ed.): Surface activity and detergency. London: Macmillan Ltd 1961
15. Hutchinson, E., Shinoda, K.: In: Solvent properties of surfactant solutions. p. 1. Shinoda, K. (ed.). New York: Dekker 1967
16. Shinoda, K.: In: Solvent properties of surfactant solutions. p. 27. Shinoda, K. (ed.). New York: Dekker 1967
17. Ekwall, P., Danielsson, I., Stenius, P.: In: Surface chemistry and colloids. MTP Int. Rev. Sci., Phys. Chem. Ser. I. vol. 7, p. 97. Kerker, M. (ed.). London: Butterworths 1972

18. Ekwall, P., Stenius, P.: In: Surface Chem. and Colloids. MTP Int. Rev. Sci., Phys. Chem. Ser. II. vol. 7, p. 215. Kerker, M. (ed.). London: Butterworths 1975
19. Hartley, G. S.: Aqueous solutions of paraffin chain salts. Paris: Hermann 1936
20. Hartley, G. S.: Ref. 5, vol. 1, p. 23
21. Ekwall, P.: Acta Acad. Abo. 4, 1 (1927)
22. Jones, E., Bury, C. R.: Philos. Mag. 4, 841 (1927)
23. Mukerjee, P.: J. Pharm. Sci. 63, 972 (1974)
24. Mukerjee, P., Cardinal, J. R.: J. Pharm. Sci. 65, 882 (1976)
25. Mukerjee, P., Mysels, K. J.: ACS Symp. Ser. 9, 239 (1975)
26. Mukerjee, P., Yang, A. Y. S.: J. Phys. Chem. 80, 1388 (1976)
27. Mukerjee, P., Mysels, K. J., Kapauan, P.: J. Phys. Chem. 71, 4166 (1967)
28. Emerson, M. F., Holtzer, A.: J. Phys. Chem. 71, 1898 (1967)
29. Gravsholt, S.: J. Colloid Interface Sci. 57, 575 (1976)
30. Hoffmann, H., Nüsslein, H., Ulbricht, W.: In: Ref. 5, vol. 1, p. 263
31. Muller, N.: In: Ref. 5, vol. 1, p. 224
32. Brun, T. S., Høiland, H., Vikingstad, E.: J. Colloid Interface Sci. 63, 89 (1978)
33. Mazer, N. A., Benedek, G. B., Carey, M. C.: J. Phys. Chem. 80, 1075 (1976)
34. Young, C. Y., et al.: J. Phys. Chem. 82, 1375 (1978)
35. Ekwall, P., Mandell, L., Solyom, P.: J. Colloid Interface Sci. 35, 519 (1971)
36. Kushner, L. M., Duncan, B. C., Hoffman, J. I.; J. Res. Natl. Bur. Stand. 49, 85 (1952)
37. Kamenka, N., Lindman, B., Brun, B.: Colloid Polym. Sci. 252, 144 (1974)
38. Danielsson, I., et al.: Progr. Colloid Polm. Sci. 61, 1 (1976)
39. Høiland, H., Vikingstad, E.: J. Colloid Interface Sci. 64, 126 (1978)
40. Vikingstad, E., Høiland, H.: J. Colloid Interface Sci. 64, 510 (1978)
41. Vikingstad, E., Skauge, A., Høiland, H.: J. Colloid Interface Sci. 66, 240 (1978)
42. Stenius, P., Ekwall, P.: Acta Chem. Scand. 21, 1643 (1967)
43. Brun, T. S., Høiland, H., Vikingstad, E.: J. Colloid Interface Sci. 63, 590 (1978)
44. Larsen, J. W., Tepley, L. B.: J. Colloid Interface Sci. 49, 113 (1974)
45. Mathews, W. K., Larsen, J. W., Pikal, M. J.: Tetrahedron Lett. 1972, 513
46. Cutler, S. G., et al.: J. C. S. Faraday I 74, 1758 (1978)
47. Elworthy, P. H., Florence, A. T.: Kolloid-Z. 208, 157 (1966)
48. Hess, L., Suranyi, L. A.: Z. Physik Chem. (Leipzig) A 184, 321 (1939)
49. Drakenberg, T., Lindman, B.: J. Colloid Interface Sci. 44, 184 (1973)
50. Persson, B. O., Drakenberg, T., Lindman, B.: J. Phys. Chem. 80, 2124 (1976)
51. Perrsson, B. O., Drakenberg, T., Lindman, B.: J. Phys. Chem., in press
52. Muller, N.: In: Reaction Kinetics in Micelles, p. 1. Cordes, E. (ed.) New York: Plenum 1973
53. Henriksson, H., Ödberg, L.: J. Colloid Interface Sci. 46, 212 (1974)
54. Ulmius, J., Lindman, B.: to be published
55. Nakagawa, T., Inoue, H., Jizomoto, H., Horiuchi, K.: Kolloid-Z. Z. Polym. 229, 159 (1969)
56. Eriksson, J. C., Gillberg, G.: Acta Chem. Scand. 20, 2019 (1966)
57. Lindman, B., et al.: In: Ref. 5, vol. 1, p. 195
58. Lindblom, G., Lindman, B., Mandell, L.: J. Colloid Interface Sci. 42, 400 (1973)
59. Gustavsson, H., Lindman, B.: J. Amer. Chem. Soc. 100, 4647 (1978)
60. Clifford, J., Pethica, B. A.: Trans. Faraday Soc. 61, 182 (1965)
61. Walker, T.: J. Colloid Interface Sci. 45, 372 (1973)
62. Robb, I. D.: J. Colloid Interface Sci. 37, 521 (1971)
63. Lindblom, G., Lindman, B., Tiddy, G. J. T.: J. Amer. Chem. Soc. 100, 2299 (1978)
64. Persson, N. O., Lindman, B.: J. Phys. Chem. 79, 1410 (1975)
65. Persson, N. O., Lindman, B.: Mol. Cryst. Liquid Cryst. 38, 327 (1977)
66. Mely, B., Charvolin, J., Keller, P.: Chem. Phys. Lipids 15, 161 (1975)
67. Henriksson, U. et al.: Chem. Phys. Lett. 52, 554 (1977)
68. Waggoner, A. S., Keith, A. D., Griffith, O. H.: J. Phys. Chem. 72, 4129 (1968)
69. Stilbs, P., Jermer, J., Lindman, B.: J. Colloid Interface Sci. 60, 232 (1977)
70. Rosenholm, J. B., Larsson, K., Dinh-Nguyen, N.: Colloid Polym. Sci. 255, 1098 (1977)
71. Okabayashi, H., Okuyama, M., Kitagawa, T.: Bull. Chem. Soc. Japan 48, 2264 (1975)

72. Rosenholm, J. B., Stenius, P., Danielsson, I.: J. Colloid Interface Sci. *57*, 551 (1976)
73. Mukerjee, P., Cardinal, J. R., Desai, N. R.: In: Ref. 5, vol. 1, p. 241
74. Gratzer, W. B., Beaven, G. H.: J. Phys. Chem. *73*, 2270 (1969)
75. Ray, A., Némethy, G.: J. Phys. Chem. *75*, 804 (1971)
76. Kalyanasundaram, K., Thomas, J. K.: In: Ref. 5, vol. 2, p. 569
77. Kalyanasundaram, K., Thomas, J. K.: J. Amer. Chem. Soc. *99*, 2039 (1977)
78. Thomas, J. K.: Acc. Chem. Res. *10*, 133 (1977)
79. Chen, M., Grätzel, M., Thomas, J. K.: J. Amer. Chem. Soc. *97*, 2052 (1975)
80. Hautula, R. R., Schore, N. E., Turro, N. J.: J. Amer. Chem. Soc. *95*, 5508 (1973)
81. Patterson, L. K., Viel, E.: J. Phys. Chem. *77*, 1191 (1973)
82. Kalyanasundaram, K., Grätzel, M., Thomas, J. K.: J. Amer. Chem. Soc. *97*, 3915 (1975)
83. Reiss-Husson, F., Luzzati, V.: J. Phys. Chem. *68*, 3504 (1964)
84. Balmbra, R. R., Clunie, J. S., Goodman, J. F.: Nature *222*, 1159 (1969)
85. Fontell, K.: In: Liquid crystals and plastic crystals. vol. 2, p. 80. Gray, G. W., Winsor, P. A. (eds.). Chichester U. K.: Horwood 1974
86. Ekwall, P., Mandell, L., Fontell, K.: Mol. Cryst. Liquid *8*, 157 (1969)
87. Klevens, H. B.: Chem. Rev. *47*, 1 (1950)
88. Swarbrick, J., Galowania, J., Bates, T. R.: J. Colloid Interface Sci. *41*, 609 (1972)
89. Jacobs, P. T., Geer R. D., Anacker, E. W.: J. Colloid Interface Sci. *39*, 611 (1972)
90. Birdi, K. S., Magnusson, T.: Colloid Polym. Sci. *254*, 1059 (1976)
91. Birdi, K. S.: In: Ref. 5, vol. 1, p. 151
92. Hartley, G. S.: J. Chem. Soc. *1938*, 1968
93. Samis, C. S., Hartley, G. S.: Trans. Faraday Soc. *34*, 1288 (1938)
94. Gerry, H. E., Jacobs, P. T., Anacker, E. W.: J. Colloid Interface Sci. *62*, 556 (1977)
95. Lindman, B., Forsén, S.: Chlorine, bromine, and iodine NMR. Physico-chemical and biological applications. Berlin, Heidelberg, New York: Springer-Verlag 1976
96. Elworthy, P. H., McDonald, C.: Kolloid-Z. Z. Polym. *195*, 16 (1964)
97. Staples, E. J., Tiddy, G. J. T.: J. C. S. Faraday I *74*, 2530 (1978)
98. Shinoda, K.: J. Colloid Interface Sci. *34*, 278 (1970)
99. Shinoda, K., Takeda, H.: J. Colloid Interface Sci. *32*, 642 (1970)
100. Friberg, S., Buraczewska, I., Ravey, J. C.: In: Ref. 5, vol. 2, p. 901
101. Friman, R., Pettersson, K., Stenius, P.: J. Colloid Interface Sci. *53*, 90 (1975)
102. Danielsson, I.: Fin. Kemistsamf. Medd. *75*, 65 (1966)
103. Danielsson, I., Stenius, P.: J. Colloid Interface Sci. *37*, 264 (1971)
104. Aniansson, E. A. G., et al.: J. Phys. Chem. *80*, 905 (1976)
105. Djavanbakht, A., Kale, K. M., Zana. R.: J. Colloid Interface Sci. *59*, 139 (1977)
106. Zana, R.: J. Phys. Chem. *82*, 2440 (1978)
107. Lindman, B., Kamenka, N., Brun, B.: J. Colloid Interface Sci. *56*, 328 (1976)
108. Lindman, B., et al.: J. Colloid Interface Sci., in press
109. Small, D. M.: Adv. Chem. Ser. *84*, 31 (1968)
110. Clifford, J., Pethica, B. A.: Trans. Faraday Soc. *60*, 216 (1964)
111. Clifford, J., Pethica, B. A.: J. Phys. Chem. *70*, 3345 (1966)
112. Fabre, H.: Thesis, Montpellier 1974
113. Becher, P.: In: Nonionic surfactants. p. 478. Schick, M. J., (ed.). New York: Dekker 1967
114. Geer, R. D., Eylar, E. H., Anacker, E. W.: J. Phys. Chem. *75*, 369 (1971)
115. Molyneux, P., Rhodes, C. T., Swarbrick, J.: Trans. Faraday Soc. *61*, 1043 (1965)
116. Evans, H. C.: J. Chem. Soc. *1956*, 579
117. Hill, T. L.: J. Chem. Phys. *36*, 153 (1962)
118. Hill, T. L.: Thermodynamics of small systems. Vol. 1 and 2. New York: Bejamin 1964
119. Hall, D. G., Pethica, B. A.: In: Nonionic surfactants, Chap. 16. Schick, M. J. (ed.) New York: Dekker 1967
120. Goodman, D. S.: J. Amer. Chem. Soc. *80*, 3887 (1958)
121. McAuliffe, C.: J. Phys. Chem. *70*, 1267 (1966)
122. Kinoshita, K., Ishikawa, H., Shinoda, K.: Bull. Chem. Soc. Japan, *30*, 1081 (1958)
123. Hermann, R. B.: J. Phys. Chem. *76*, 2754 (1972)

124. Amidon, G. L. et al.: J. Phys. Chem. *79*, 2239 (1976)
125. Gill, S. J., Wadsö, I.: Proc. Natl. Acad. Sci. USA *73*, 2955 (1976)
126. Ref. 4. Chap. 4
127. Bohon, R. L., Claussen, W. F.: J. Amer. Chem. Soc. *73*, 1571 (1951)
128. Gill, S. J., Nichols, N. F., Wadsö, I.: J. Chem. Thermodyn. *8*, 445 (1976)
129. Nichols, N., et al.: J. Chem. Thermodyn. *8*, 1081 (1976)
130. Franks, F., Ives, D. J. G.: Quart. Rev. Chem. Soc. *20*, 1 (1966)
131. Høiland, H., Vikingstad, E.: Acta Chem. Scand. *A 30*, 182 (1976)
132. Hertz, H. G.: In: Progress in NMR spectroscopy. Vol. III. p. 159. Oxford: Pergamon 1967
133. Davidson, D. W.: In: Water. A comprehensive treatise. Vol. 2. Chap. 3. Franks. F. (ed.). New York. Plenum 1973
134. Narten, A. H., Lindenbaum, S.: J. Chem. Phys. *51*, 1108 (1969)
135. Franks, F. (ed.): Water. A comprehensive treatise. Vol. 1–5. New York. Plenum 1972–76
136. Ben-Naim, A.: Water and aqueous solutions. New York: Plenum 1974
137. Marcelja, S., et al.: J. C. S. Faraday II *73*, 630 (1977)
138. Owicki, J. C., Scheraga, H. A.: J. Amer. Chem. Soc. 99, 7413 (1977)
139. Swaminathan, S., Harrison, S. W., Beveridge, D. L.: J. Amer. Chem. Soc. *100*, 5705 (1978)
140. Holtzer, A., Holtzer, M. F.: J. Phys. Chem. *78*, 1442 (1974)
141. Benjamin, L.: J. Colloid Interface Sci. *22*, 386 (1966)
142. Benjamin, L.: J. Phys. Chem. *70*, 3790 (1966)
143. Goddard, E. D., Benson, G. C.: Can. J. Chem. *35*, 986 (1957)
144. Pilcher, G., et al.: J. Chem. Thermodyn. *1*, 381 (1969)
145. Clint, J. H., Sumida, K.: Chem. Pharm. Bull (Japan) *22*, 1108 (1974)
146. Paredes, S. et al.: Colloid Polym. Sci. *254*, 637 (1976)
147. Kishimoto, H., Sumida, K.: Chem. Pharm. Bull (Japan) *22*, 1108 (1974)
148. Goddard, E. D., Hoeve, C. A. J., Benson, G. C.: J. Phys. Chem. *61*, 593 (1957)
149. Benjamin, L.: J. Phys. Chem. *68*, 3575 (1964)
150. Debye, P.: J. Phys. Chem. *53*, 1 (1949)
151. Debye, P.: Ann. N. Y. Acad. Sci. *51*, 573 (1949)
152. Reich, I.: J. Phys. Chem. *60*, 257 (1956)
153. Hoeve, C. A. J., Benson, G. C.: J. Phys. Chem. *61*, 1149 (1957)
154. Poland, D. C., Scheraga, H. A.: J. Phys. Chem. *69*, 2431 (1965)
155. Poland, D. C., Scheraga, H. A.: J. Phys. Chem. *69*, 4425 (1965)
156. Poland, D. C., Scheraga, H. A.: J. Colloid Interface Sci. *21*, 273 (1966)
157. Nemethy, G., Scheraga, H. A.: J. Phys. Chem. *66*, 2888 (1962)
158. Tanford, C.: J. Mol. Biol. *67*, 59 (1972)
159. Tanford, C.: Proc. Nat. Acad. Sci. *71*, 1811 (1974)
160. Tanford, C.: J. Phys. Chem. *78*, 2469 (1974)
161. Tanford, C.: In: Ref. 5, Vol. 1, p. 119
162. Ruckenstein, E., Nagarajan, R.: J. Phys. Chem. *79*, 2622 (1975)
163. Nagarajan, R., Ruckenstein, E.: J. Colloid Interface Sci. *60*, 221 (1977)
164. Ruckenstein, E., Nagarajan, R.: In: Ref. 5, Vol. 1 p. 133
165. Israelachvili, J. N., Mitchell, J. J., Ninham, B. N.: J. C. S. Faraday II, *72*, 1525 (1976)
166. Eriksson, F., Eriksson, J. C., Stenius, P.: Presented at the National Colloid Sci. Meeting Knoxville 1978
167. Mazer, N. A., Carey, M. C., Benedek, G. B.: In: Ref. 5. Vol. 1, p. 359
168. Winsor, P. A.: Chem. Rev. *68*, 1 (1968)
169. Götz, K. G., Heckman, K.: Z. Physik. Chem. *20*, 42 (1959)
170. Ulmius, J., Wennerström, H.: J. Magn. Resonance *28*, 309 (1977)
171. Wennerström, H.: J. Colloid Interface Sci. *68*, 589 (1979)
172. Ulmius, J. et al.: J. Phys. Chem. *83*, 2232 (1979)
173. Tartar, H. V.: J. Phys. Chem. *59*, 1195 (1955)
174. Tanford, C., Nozaki, Y., Rodhe, M.: J. Phys. Chem. *81*, 1555 (1977)
175. Berne, B. J., Pecora, R.: Dynamic Light Scattering. With Applications to Chemistry, Biology and Physics. New York: Wiley 1976

176. Tiddy, G. J. T.: Personal communication
177. Ref. 2. Chap. 1
178. Corkill, J. M., Goodman, J. F., Tate, J. R.: Trans. Faraday Soc. 60, 986 (1964)
179. Mazer, N. A., et al.: Ref. 5. Vol 1 p. 383
180. Barker, C. A., et al.: J. C. S. Faraday I 70, 154 (1974)
181. Hall, D. G.: Trans. Faraday. Sci. 60, 1359 (1970)
182. Mukerjee, P.: J. Pharm. Sci. 60, 1531 (1971)
183. Benjamin, L.: J. Phys. Chem. 70, 3790 (1966)
184. Larsen, J. W., Magid, L. J.: J. Phys. Chem. 78, 834 (1974)
185. Hakala, M. R., Rosenholm, J. B., Stenius, P.: submitted
186. Williams, R. J., Phillips, J. N., Mysels, K. J.: Trans. Faraday Soc. 51, 728 (1955)
187. Aniansson, E. A. G.: Ber. Bunsenges. Phys. Chem. 82, 981 (1978)
188. Hoffmann, H.: Ber. Bunsenges. Phys. Chem. 82, 988 (1978)
189. Anderson, J. L., Reed, C. C.: J. Chem. Phys. 64, 3240 (1976)
190. Phillies, G. D. J., Benedek, G. B., Mazer, N. A.: J. Chem. Phys. 65, 1883 (1976)
191. Anderson, J. L., Rauh, F., Morales, A.: J. Phys. Chem. 82, 608 (1978)
192. Götz, K. G., Heckmann, K.: J. Colloid Sci. 13, 206 (1958)
193. Scheraga, H. A., Backus, J. K.: J. Amer. Chem. Soc. 73, 5108 (1951)
194. Ulmius, J., Wennerström, H.: J. Magn. Resonance 28, 309 (1977)
195. Henriksson, U., et al.: J. Phys. Chem. 81, 76 (1977)
196. Johansson, L., Lindblom, G., Nordén, B.: Chem. Phys. Lett. 39, 128 (1976)
197. Kamenka, N., et al.: J. Chim. Phys. 74, 510 (1977) unpublished
198. Gravsholt, S.: In: Proc. VIth Int. Congress on Surface Active Substances, Zurich, 1972,
     vol. II, p. 805. München: Hanser 1973
199. Johansson, L. B. Å., et al.: Ber. Bunsenges. Phys. Chem. 82, 978 (1978)
200. Mukerjee, P.: J. Phys. Chem. 76, 565 (1972)
201. Mukerjee, P.: In: Ref. 5, vol. 1, p. 171
202. Rohde, A., Sackmann, E.: Ber. Bunsenges. Phys. Chem. 82, 978 (1978)
203. Turro, N. J., Yekta, A.: J. Amer. Chem. Soc. 100, 5951 (1978)
204. Granath, K.: Acta Chem. Scand. 7, 297 (1953)
205. Mysels, K. J., Princen, L. H.: J. Phys. Chem. 63, 1696 (1959)
206. Chapman, D., Williams, R. M., Ladbrooke, B. D.: Chem. Phys. Lipids 1, 445 (1967)
207. Wennerström, H.: Chem. Phys. Lett. 18, 41 (1973)
208. Seelig, J.: Quart. Rev. Biophys. 10, 353 (1977)
209. Mautsch, H. H., Saito, H., Smith, I. C. P.: Prog. NMR Spectrosc. 11, 211 (1977)
210. Schara, M., Pašnik, F., Šentjurc, M.: Croat. Chem. Acta 48, 147 (1976)
211. Harkins, W. D., Matton, R. W., Corrin, M. L.: J. Colloid Sci. 7, 105 (1946)
212. Shinoda, K., Hildebrand, J. H.: J. Phys. Chem. 62, 295 (1958)
213. Mukerjee, P.: Kolloid – Z. Z. Polym. 236, 76 (1970)
214. Shinitzsky, M., et al.: Biochemistry 10, 2106 (1971)
215. Waggoner, A. S., Keith, A. D., Griffith, O. H.: J. Phys. Chem. 72, 4129 (1968)
216. Ulmius, J., et al.: J. Colloid Interface Sci. 65, 88 (1978)
217. Rosenholm, B., Drakenberg, T., Lindman, B.: J. Colloid Interface Sci. 63, 538 (1978)
218. Wennerström, H., et al.: J. Amer. Chem. Soc., in press
219. Lindblom, G., Wennerström, H.: Biophys. Chem. 6, 167 (1977)
220. Lindman, B., Brun, B.: J. Colloid Interface Sci. 42, 388 (1973)
221. Kamenka, N., Brun, B., Lindman, B.: Presented at the VII Int. Congr. Surf. Active Subst.,
     Moscow 1976
222. Kamenka, N., et al.: C. R. Acad. Sci. Paris 284, 403 (1977)
223. Kale, K. M., Zana, R.: J. Colloid Interface Sci. 61, 312 (1977)
224. Söderman, O.: personal communication
225. Gustavsson, H., et al.: In: Liquid crystals and ordered fluids, vol. 2. p. 162. Johnson, J. F.,
     Porter, R. S. (eds.) New York: Plenum 1974
226. Gustavsson, H., Lindman, B.: J. Amer. Chem. Soc. 97, 3923 (1975)
227. Anacker, E. W., Ghose, H. M.: J. Amer. Chem. Soc. 90, 3161 (1968)

228. Clifford, J., Pethica, B.A.: Trans. Faraday Soc. *60*, 1483 (1964)
229. Aniansson, E. A. G.: J. Phys. Chem. *82*, 2805 (1978)
230. Mukerjee, P.: J. Colloid Sci. *19*, 722 (1964)
231. Ekwall, P., Holmberg, P.: Acta Chem. Scand. *19*, 455 (1965)
232. Courchene, W. L.: J. Phys. Chem. *68*, 1870 (1964)
233. Tokiwa, F., Ohki, K.: J. Phys. Chem. *71*, 1343 (1967)
234. Puyal, M.C.: D. E. A., Faculté des Sciences, Montpellier, 1978
235. Stigter, D.: J. Phys. Chem. *78*, 2480 (1974)
236. Muller, N.: J. Magn. Reson. *23*, 203 (1977)
237. Menger, F. M., Jerkunica, J. M., Johnston, J. C.: J. Amer. Chem. Soc. *100*, 4676 (1978)
238. Podo, F., Ray, A., Némethy, G.: J. Amer. Chem. Soc. *95*, 6164 (1973)
239. Clemett, C. J.: J. Chem. Soc. A *1970*, 2251
240. Corkill, J. M., Goodman, J. F., Wyer, J.: Trans. Faraday Soc. *65*, 9 (1969)
241. Tiddy, G. J. T., Walsh, M. F., Wyn-Jones, E.: Chem. Commun. *1979*, 252
242. Mukerjee, P.: J. Phys. Chem. *66*, 943 (1962)
243. Lindman, B., Ekwall, P.: Mol. Cryst. *5*, 79 (1968)
244. Ekwall, P., Mandell, L.: Acta Chem. Scand. *22*, 699 (1968)
245. Ray. A., Mukerjee, P.: J. Phys. Chem. *70*, 2138 (1966)
246. Mukerjee, P., Ray, A.: J. Phys. Chem. *70*, 2144 (1966)
247. Mukerjee, P., Ray, A.: J. Phys. Chem. *70*, 2150 (1966)
248. Lindblom, G., Lindman, B.: J. Phys. Chem. *77*, 2531 (1973)
249. Stigter, D.: J. Phys. Chem. *68*, 3603 (1964)
250. Andrasko, J., Forsén, S.: Biochem. Biophys. Res. Commun. *60*, 813 (1974)
251. Kamenka, N., Fabre, H., Lindman, B.: C. R. Acad. Sci. Paris *281*, 1045 (1975) unpublished
252. Cardinal, J. R., Mukerjee, P.: J. Phys. Chem. *82*, 1614 (1978)
253. Mukerjee, P., Cardinal, J. R.: J. Phys. Chem. *82*, 1620 (1978)
254. Rosenholm, J. B., Lindman, B.: J. Colloid Interface Sci. *57*, 362 (1976)
255. Chorro, M.: Thesis, Faculté des Sciences, Montpellier, 1977
256. Fabre, H.: Thesis, Faculté des Sciences, Montpellier, 1976
257. Aniansson, E. A. G., Wall, S. N.: J. Phys. Chem. *78*, 1024 (1974)
258. Aniansson, E. A. G., Wall, S. N.: J. Phys. Chem. *79*, 857 (1975)
259. Aniansson, E. A. G., Wall, S. N.: In: Chemical and Biological Applications of Relaxation Spectrometry. p. 223. Wyn-Jones, E. (ed.) Dordrecht: Reidel 1975
260. Aniansson, E. A. G., Almgren, M., Wall, S. N.: In: Chemical and Biological Applications of Relaxation Spectrometry. p. 239. Wyn-Jones, E. (ed.). Dordrecht: Reidel 1975
261. Wall, S. N.: Thesis, University of Gothenburg, 1977
262. Almgren, M., Aniansson, E. A. G., Holmåker, K.: Chem. Phys. *19*, 1 (1977)
263. Yasunaga, T., Fujii, S., Miura, M.: J. Colloid Interface Sci. *30*, 399 (1969)
264. Zana, R., Lang, J.: Comptes Rendus Acad. Sci. (Paris) *266*, 1347 (1968)
265. Rassing, J., Sams, P. J., Wyn-Jones, E.: J. Chem. Soc. Faraday II *70*, 1247 (1974)
266. Lang, J., Auborn, J. J., Eyring, E. M.: J. Colloid Interface Sci. *41*, 484 (1972)
267. Yasunaga, T., Takeda, K., Harada, S.: J. Colloid Interface Sci. *42*, 457 (1973)
268. Mijnlieff, P. F., Ditmarsch, R.: Nature *208*, 889 (1965)
269. Takeda, K., Yasunaga, T.: J. Colloid Interface Sci. *40*, 127 (1972)
270. Herrmann, U., Kahlweit, M.: Ber. Bunsenges. Phys. Chem. *77*, 1119 (1973)
271. Janjic, T., Hoffmann, H.: Z. Phys. Chem. Neue Folge *86*, 322 (1973)
272. Kresheck, G. C., et al.: J. Amer. Chem. Soc. *88*, 246 (1966)
273. Bennion, B. C., et al.: J. Phys. Chem. *73*, 3288 (1969)
274. Takeda, K., Yasunaga, T.: J. Colloid Interface Sci. *45*, 406 (1973)
275. Tondre, C., Lang, J., Zana, R.: J. Colloid Interface Sci. *45*, 406 (1973)
276. Lang, J., et al.: J. Phys. Chem. *79*, 276 (1975)
277. Ref. 104. Footnote 6
278. Oakes, J.: J. Chem. Soc. Faraday II *68*, 1464 (1972)
279. Thomas, J. K., Grieser, F., Wong, M.: Ber. Bunsenges. Phys. Chem. *82*, 937 (1978)
280. Grünhagen, H. H.: J. Colloid Interface Sci. *53*, 282 (1975)

281. Cordes, E. H. (ed.).: Reaction Kinetics in Micelles, New York: Plenum 1973
282. Martinek, K., et al.: In: Ref. 5, Vol. 2, p. 489
283. Bruice, T. C.: In: The Enzymes. Boyer, P. D. Vol. 2, chap. 4. (ed.). New York: Academic Press 1970
284. Romsted, L. S.: In: Ref. 5. Vol. 2, p. 509.
285. Diekmann, S., Frahm, J.: Ber. Bunsenges. Phys. Chem. *82*, 1013 (1978)
286. Diekmann, S., Frahm, J.: Submitted for publication
287. Stigter, D., Williams, R. J., Mysels, K. J.: J. Phys. Chem. *59*, 1330 (1955)
288. Stejskal, E. O., Tanner, J. E.: J. Chem. Phys. *49*, 288 (1965)
289. Bull, T., Lindman, B.: Mol. Cryst. Liquid Cryst. *28*, 155 (1974)
290. Mazo, R. M.: J. Chem. Phys. *43*, 2873 (1965)
291. Ref. 175. Chap. 9
292. Lie, G. C., Clementi, E., Yoshimine, M.: J. Chem. Phys. *64*, 2314 (1976)
293. Mezei, M., Swaminathan, S., Beveridge, D. L.: J. Am. Chem. Soc. *100*, 3255 (1978)
294. Clementi, E., et al.: FEBS Lett. *100*, 313 (1979)
295. Shinoda, K., Hutchinson, E.: J. Phys. Chem. *60*, 577 (1962)
296. Hoskin, N. R.: Trans. Faraday Soc. *49*, 1471 (1953)
297. Loeb, A. L., Overbeek, J. Th. G., Wiersems, P. H.: The electrical double layer around a spherical colloid particle. Cambridge, MA.: MIT 1961
298. Bell, G. M., Dunning, A. J.: Trans. Faraday Soc. *66*, 500 (1970)
299. Mille, M., Vanderkooi, G.: J. Colloid Interface Sci. *59*, 211 (1977)
300. Gunnarsson, G., Wennerström, H.: Submitted
301. Bell, G. M., Levine, S.: In: Chemical Physics of ionic solutions. p. 409. Conway, B. E., Barradas, R. G. (eds.). New York: Wiley 1966
302. Marcus, R. A.: J. Chem. Phys. *23*, 1057 (1955)
303. Overbeek, J. Th. G., Stigter, D.: Rec. Trav. Chim. *75*, 1263 (1956)
304. Gunnarsson, G., Olofsson, G., Wennerström, H., Zacharov, A.: J. C. S. Faraday 1, in press
305. Stigter, D.: J. Colloid Interface Sci. *47*, 473 (1974)
306. Oosawa, F.: Polyelectrolytes. New York: Marcel Dekker 1971
307. Manning, G. S.: Ann. Rev. Phys. Chem. *23*, 117 (1972)
308. Engström, S., Wennerström, H.: J. Phys. Chem. *82*, 2711 (1978)
309. Jönsson, B., Wennerström, H.: Chem. Scripta, in press
310. Guéron, M., Weisbuch, G.: Submitted
311. Mysels, K. J., Princen, L. H.: J. Phys. Chem. *69*, 4038 (1965)
312. Sasaki, T., et al.: Bull. Chem. Soc. Jap. *48*, 1397 (1975)
313. Huff, H., McBain, J. W., Brady, A. P.: J. Phys. Chem. *55*, 311 (1950)
314. Fuoss, R. M., Katchalsky, A., Lifson, S.: Proc. Natl. Acad. Sci: U. S. *37*, 579 (1951)
315. Kiessig, H., Philipoff, W.: Naturwiss, *27*, 593 (1939)
316. Hess, K., Gundermann, J.: Ber. *70*, 1800 (1946)
317. Hartley, G. S.: Nature *163*, 787 (1949)

Received April 9, 1979

Surfactants in Nonpolar Solvents
Aggregation and Micellization

Hans-Friedrich Eicke

Institute for Physical Chemistry, University of Basel, CH-4056 Basel, Switzerland

Table of Contents

1 Introduction

Since about 1950 a continuously increasing interest can be observed with regard to nonpolar detergent solutions. In the last few years, however, the wealth of information due to new and fascinating properties and applications of "reversed" or "inverted" micellar aggregates has become so extensive that an independent presentation of surfactant systems in nonpolar solvents is certainly justified, particularly because the physico-chemical properties of hydrophilic and lipophilic micelles are in many respects remarkably different. The following considerations will accordingly be concerned with aggregational phenomena characteristic of surface active substances, i.e., surfactants or detergents, in nonpolar media. The tendency to be surface active and to form (eventually) spontaneously aggregates which are in a thermodynamical equilibrium with monomers or molecular subunits (oligomers) is intrinsically related to the typical surfactant structure: it appears that two spatially separated antagonistic molecular moieties are essential which are each uniform in such a sense as to being predominantly ionic or polar, i.e., hydrophilic (for example, suitable to form hydrogen bonds) or apolar or lipophilic, i.e., consisting of hydrocarbon groups ("tails").

As in aqueous detergent solutions it has become customary to distinguish nonionic and ionic surfactants, thereby subdividing the latter into cationic and anionic surfactants. Some authors also include amphoteric surfactants in this group. Typical representatives of these surfactant types are given in Table 1. The distinction between

Table 1. Structure of Surfactants (from: Nerdel, F.: Organische Chemie Berlin: De Gruyter 1964)

Type	Hydrophilic Moiety	Examples	
Anionic	$-COO^-$	$CH_3-(CH_2)_a-COO^-$	Na^+
	$-O-SO_2-O^-$	$CH_3-(CH_2)_a-O-SO_2-O^-$	Na^+
	$-SO_2-O^-$	$CH_3-(CH_2)_a-SO_2-O^-$	Na^+
		$CH_3-(CH_2)_b-\langle\rangle-SO_2--O^-$	Na^+
	$-CO-N-CH_2-CH_2-SO_2-O^-$ $\quad\quad\;\; R$	$CH_3-(CH_2)_a-CO-N-CH_2-CH_2-SO_2-O^-$ $\quad\quad\quad\quad\quad\quad\quad\quad\; R$	Na^+
Cationic	$\begin{array}{c}CH_3\\ \mid\\ -N^+-CH_3\\ \mid\\ CH_3\end{array}$	$\begin{array}{c}CH_3\\ \mid\\ CH_3-(CH_2)_a-N^+-CH_3\\ \mid\\ CH_3\end{array}$	Cl^-
Amphoteric	$\begin{array}{c}CH_3\\ \mid\\ -N^+-CH_2-CO-O^-\\ \mid\\ CH_3\end{array}$	$\begin{array}{c}CH_3\\ \mid\\ CH_3-(CH_2)_a-N^+-CH_2-CO-O^-\\ \mid\\ CH_3\end{array}$	
Nonionic	$-O-(CH_2-CH_2-O-)_c-H$	$CH_3-(CH_2)_a-O-(CH_2-CH_2-O-)_c-H$	
		$CH_3-(CH_2)_b-\langle\rangle-O-(CH_2-CH_2-O-)_c-H$	
	$\begin{array}{c}CH_2OH\\ \mid\\ -CO-NH-C-CH_2OH\\ \mid\\ CH_2OH\end{array}$	$\begin{array}{c}CH_2OH\\ \mid\\ CH_3-(CH_2)_a-CO-NH-C-CH_2OH\\ \mid\\ CH_2OH\end{array}$	

a=10—20, b=8—16, c=5—20

cationic and anionic surfactants is somewhat formal and refers conventionally to the (limiting) case where a relatively small counterion is attached to a large "lipophilic" anion or cation, i.e., the notation refers to the ion which is bound covalently to the hydrocarbon moiety of the surfactant (see examples in Table 1). Apart from this more conventional definition there are, actually, some physico-chemical differences which make such a distinction meaningful. A detailed discussion will be given in the following paragraphs.

Nonionic detergents might well be characterized as being heterogeneous material, particularly if the hydrophilic moiety is based on the addition of ethylene oxide[11]. Molecular uniformity is, therefore, not easy to obtain with these compounds. Due to the possibility of a delicate variation of the so-called hydrophile-lipophile balance[212], these systems prompted considerable industrial interest.

Regarding the nonaqueous solvents used in the pertinent investigations they can be classified according to the following scheme[173, 174]:

(a) those solvents in which inverted (reversed) micelles are formed, and

(b) those in which micelles do not exist at low and medium surfactant concentration.

The following discussion deals exclusively with type (a), i.e., inverted micellar aggregates in nonpolar hydrocarbon solvents. Thus, simple aliphatic, cycloaliphatic and aromatic hydrocarbons were generally selected, also carbon tetrachloride has been considered to be favorable. The more apolar the solvent the more pronounced aggregation tendency of the surfactants has been observed. Hence, solvent effects on detergent aggregation were soon realized. Particularly, solvents which are hydrogen bond donors or acceptors, even if they appear to belong to the class of nonpolar solvents like dioxane[39] or ethylacetate show strikingly disaggregating effects on micellar aggregates, or they prevent extensive association. Examining the available experimental data on nonpolar, aprotic, surfactant solutions which have been summarized in a number of reviews[197, 190, 82, 83, 119, 11, 71] the formation of so-called reversed (or inverted) micelles in these media is generally considered to be beyond doubt. The notions were adopted from the corresponding observations in aqueous detergent solutions where already sufficient research experience existed over a long period of time. From the IUPAC Information Bulletin[102] the most appropriate hints to define a micellar aggregate can be obtained. The obvious difference between micelles in polar (generally aqueous) and nonpolar surfactant solutions which had soon been realized is their mutual structural reversion. Consequently, one speaks of so-called reversed or inverted micelles in a nonpolar environment. Thus, the inverted micelle is visualized to be built up by a polar core covered by hydrocarbon tails of the respective surfactant molecules.

A remarkable feature of the inverted micelles are their, in general, moderate aggregation numbers which contrast the large micellar aggregates in aqueous surfactant solutions. (A list of aggregation numbers of various micelles and (gel) forming compounds is given in Table 2a, b according to Singleterry[197]). This fact would imply that the onset of micelle formation according to the mass action law should be less pronounced in many instances of nonpolar detergent solutions. This has been often observed, in particular, with the cationic surfactant (see, for example[113]), and caused much uncertainty regarding the determination of a *critical micelle concen-*

Table 2a. Aggregation and micelle formation in hydrocarbon solvents

Solute	Solvent	Temp. (°C)	Solute concentration	Aggreg. no.	Ref.
Tetraisoamylammonium thiocyanate	benzene	5.4	0.02 –0.42 molal	10–25	26)
n-Octadecyltri-n-butylammonium formate	benzene	5.4	0.001–0.0033 molal	10–22	26)
n-Amyltri-n-butylammonium iodide	benzene	5.4	0.001–0.50 molal	3–22	26)
Tetra-n-butylammonium perchlorate	benzene	5.4	0.001–0.014 molal	3–6	179)
Sodium bis(2-ethylhexyl) sulfosuccinate	dodecane	30	1 g/100 g	(32)	142)
		30	$1 + 0.2\ H_2O$ g/100 g	(28)	142)
		30	$1 + 1.0\ H_2O$ g/100 g	(350)	142)
Sodium dinonylnaphthalene sulfonate	benzene	25	$10^{-5}-10^{-3}$ molar	12	103)
Barium dinonylnaphthalene sulfonate	benzene	25	$10^{-5}-10^{-3}$ molar	7	103)
Barium petroleum sulfonate	benzene	25	0.001 molar	34	103)
Calcium cetylphosphate + calcium alkylphenolate	lube oil	25	0.013 molal	(20–30)	163)
Hexanolamine oleate	benzene	5.4	0.089–0.16 molal	3+	87)
Hexanolamine caprylate	benzene	5.4	0.072 molal	4+	87)
Dodecylamine oleate	cyclohexane	6.5	1,0 g/100 g	5	160)
Dodecylamine propionate	cyclohexane	6.5	1.0 g/100 g	10	160)
Nonaethyleneglycol monolaurate	cyclohexane + water	6.5	0.101 molal	4+	87)
Alkyd resin	heptane	25	0.35–3 g/100 ml	(~100)	187)

Table 2b. Aggregation and micelle formation by metal carboxylates in hydrocarbon solvents

Solute	Solvent	Temp. (°C)	Solute concentration	Aggreg. no.	Ref.
Zinc laurate	toluene	111	0.6–1.9 g/100 ml	5–6	156)
Magnesium laurate	toluene	111	0.5–2.3 g/100 ml	6–33	156)
Copper laurate	toluene	111	0.5–1.5 g/100 ml	5–18	156)
Ferric trilaurate	toluene	111	0.005–0.05 molar	2	156)
Zinc octanoate	toluene	111	Extrapolated to ∞ dilution	6.3	156)
Zinc deconoate	toluene	111	Extrapolated to ∞ dilution	5.4	156)
Zinc laurate	toluene	111	Extrapolated to ∞ dilution	4.8	156)
Zinc myristate	toluene	111	Extrapolated to ∞ dilution	4.2	156)
Zinc stearate	toluene	111	Extrapolated to ∞ dilution	3.2	156)
Sodium phenylstearate + 0.2 mole phenyl-stearic acid per mole of soap	benzene	25	1.73 g/100 ml	200	92)
Sodium phenylstearate + trace of water	benzene	25	0.68 g/100 ml	(5000)	92)
Calcium xenylstearate + 2 moles water per mole of soap	benzene	25	10^{-6}–10^{-3} molar	24	196)
Calcium xenylstearate + trace of water	benzene	25	3.4 g/100 ml	(1000)	6)
Aluminum dicaprylate	benzene	30	0.12 g/100 ml	(640)	188)
Aluminum dicaprylate	benzene	30	0.30 g/100 ml	(880)	188)
Aluminum dilaurate	benzene	30	0.10 g/100 ml	(840)	188)
Aluminum dilaurate	benzene	30	0.40 g/100 ml	(1330)	188)
Aluminum dimyristate	benzene	30	0.075 g/100 ml	(520)	188)
Aluminum dimyristate	benzene	30	0.30 g/100 ml	(980)	188)
Aluminum dipalmitate	benzene	30	0.10 g/100 ml	(600)	188)
Aluminum dipalmitate	benzene	30	0.30 g/100 ml	(700)	188)
Aluminum distearate	benzene	30	0.20 g/100 ml	(670)	188)
Aluminum distearate	benzene	30	0.30 g/100 ml	(970)	188)

tration (CMC). The occurrence of a larger manifold of aggregational patterns (compared with aqueous systems) as encountered with detergents in nonpolar solutions makes it even more difficult to fix a critical micelle concentration. This urges one to be cautious in determining and evaluating so-called apparent CMC-values. The often insignificant changes of the physical parameters used to follow the aggregational process simply forbid a reliable statement concerning a critical concentration. It has been suggested[47], accordingly, instead of looking for the onset of an aggregational process to utilize, primarily, other criteria which are suitable to conclude a possible micellization: for example, the concentration independence of the aggregate size within the accuracy of the employed method. The number or weight average molecular weights obtained from any suitable experiment are to be analyzed according to the "mass action" or "multiple equilibrium" models both considered to be limiting cases of an aggregational process. The degree of coincidence between a real system and the models has to be checked by a careful analysis. An experimental fit corresponding to one of the two models would then be considered to represent a preselection in favor or against a principally existing CMC.

A particular problem encountered with nonpolar detergent solutions is the lack of suitable methods to determine critical micelle concentrations. This is due mainly to two reasons:

(i) A number of techniques are exclusively applicable in aqueous systems, for example, electrochemical and many of the standard relaxation methods (temperature and pressure jump techniques). Conductivity measurements have to be considered more as an exception than a rule since they can be applied only in those cases where the critical concentration is not too small, i.e., $10^{-3} - 10^{-2}$ mol dm^{-3}; many examples were reported of nonpolar surfactant solutions where the CMC-region is considerably smaller. More elaborate techniques have then to be applied in order to determine electrical conductivities[38].

(ii) Even those methods which would be principally adapted to nonpolar detergent solutions, suffer frequently from the insufficient sensitivity at low CMC-values. Dielectric techniques are probably most promising[53] and a relative new method, recently introduced to the study of nonpolar surfactant systems, i.e., the positron annihilation technique[101]. These methods aim at a least possible disturbance of the investigated systems. Another, quite successful technique, was suggested by Kaufman and Singleterry, namely the addition of a suitable fluorescence indicator to the detergent solutions. Possible objections to this procedure will be discussed later.

A list of CMC values referring to a number of typical ionic surfactants in nonpolar solvents has been published by Nakagawa and Shinoda[155]. A collection of CMCs and mean aggregation numbers of one of the most extensively investigated anionic surfactants, i.e., AOT in different nonpolar organic solvents is shown in Table 3 (Magid[139]). A phenomenon encountered with the determination of the solubility of surfactants in aqueous and nonpolar solvents which is closely related to the critical micelle concentration is the Krafft point. Due to the low solubilities of many surfactants in nonpolar media the critical concentration where aggregates start to form, in particular micellar aggregates, is higher than the equilibrium concentration of the surfactant at the specified temperature. Thus, raising the temperature increases the equilibrium concentration (= solubility) of the monomers up to the Krafft point,

Table 3. Mean aggregation numbers ($\bar{n}$) and CMCs of aerosol OT (AOT) in different hydrocarbon solvents (from [139])

Solvent	Temperature (°C)	Technique[a]	CMC, M	$\bar{n}$	Ref.
Cyclohexane	37	VPO	$3.9\ 10^{-4}$	17	210)
	28 ± 4	LS	$1.3\ 10^{-3}$	56	117)
	25	LS	–	39	77)
	40	VPO	–	18.2	123)
Benzene	37	VPO	$3.5\ 10^{-4}$	13	210)
	37	WS	–	18	114)
	RT	PA	$2.2\ 10^{-3}$	–	101)
	28 ± 4	LS	$2.7\ 10^{-3}$	23.6	117)
	20	TCNQ	$2.0\ 10^{-3}$	–	153)
	25	VPO, LS	–	15	38)
CCl$_4$	25	VPO	$1.6\ 10^{-4}$	17	210)
	37	VPO	$4.0\ 10^{-4}$	17	210)
	20	TCNQ	$6.0\ 10^{-4}$	–	153)
	25	LS	–	17	123)
	40	VPO	–	20.6	123)
CDCl$_3$	30	NMR	–	5	211)
2,2,4-Trimethyl pentane	25	VIS	–	220	162)
	40	VPO	–	21	123)
	25	LS	$4.9\ 10^{-4}$	15	53)
n-octane	25	LS	–	30	77)
n-decane	25	LS	–	37	77)
n-dodecane	25	LS	–	44	77)
n-dodecane	25	UC	–	28	162)

[a] VPO: vapour pressure osmometry; LS: Light scattering;
WS: water solubilization; PA: positron annihilation;
TCNQ: solubilization of 7,7,8,8-tetracyanoquinodimethane;
VIS: viscosity; UC: ultracentrifugation.

i.e., critical concentration. Above this point, the total solubility is determined by the aggregates. Thermodynamically, this behavior is to be understood from the structure of the mass action law and the fact that the chemical potential of the particles in the solution is essentially determined by the chemical potential of the monomeric surfactant molecules which changes only slowly with the total weighed-in concentration above the critical concentration[190, 202]. A detailed discussion regarding the relation between structure and solubility is given by Kertes[113].

The micellar phenomenon cannot be discussed without considering a surfactant property which is intrinsically related to the very existence of micelles: the so-called *detergency,* i.e., the ability of surfactant molecules to take up (= solubilize) polar material, for example, water in the polar core of the inverted micelles. Thus, micellization and solubilization are competitive processes. It is obvious that the tendency to solubilize minute amounts of polar impurities, particularly, water is quite pronounced. In principle it must appear, therefore, doubtful whether it is reasonable at all to discuss true binary systems, i.e., surfactant plus solvent, except by way of an extrapolation. It could be stated, accordingly, that a "true" CMC referring to a binary system cannot be determined. Hence, it is a matter of convention to define a suitable reference state. Thus, any solubilized probe molecule has to be considered to produce a shift of the CMC. This solubilization of an additional component will have to be discussed later on, since a well-known method used to determine CMC-values utilizes such a third component as an indicator. The possibility to solubilize considerable amounts of, for example, water, i.e., to produce water in oil (= W/O) microemulsions, which are thermodynamically stable emulsions of water in oil in the presence of surfactant molecules has been most actively investigated by K. Shinoda and S. Friberg together with their coworkers, see, e.g.[195]. This research opened up a large area of applications including photochemical reactions and, generally, catalyzed organic reactions in microemulsions, polymerization processes and pharmaceutical applications.

More recently the interest centered around the remarkable *catalytic activity* of microemulsions produced by cationic surfactants. These systems have been extensively described by Fendler and Fendler[69, 71, 74], also with respect to biological relevant model systems.

Increasing the water content the W/O-microemulsion merges into a O/W-microemulsion. Increasing the amount of surfactant leads, eventually, to the formation of liquid crystalline (200) phases usually represented in ternary phase diagrams which have been most successfully investigated by Ekwall and his school for many years in Åbo and Stockholm[63, 65, 66]. This research has been continued by Shinoda and Friberg (see e.g. [195]). The phenomena encountered with three or even more component systems have to be considered as a generalization with respect to an increase in the manifold of aggregational patterns comprising the micellization as a special association phenomenon.

2 Aggregation and Micelle Formation

2.1 Thermodynamic Models

As in aqueous surfactant solutions there exists an aggregational tendency of amphiphilic molecules in nonaqueous aprotic, i.e., nonpolar, solvents. It is generally recognized that the role of the various interactions governing the formation of detergent aggregates in nonpolar media differs from that in aqueous solutions, in spite of the apparent similar building principle of lipophilic and hydrophilic aggregates or micelles.

Since thermodynamics is not concerned with the particular intermolecular interactions, thermodynamic models describe association processes, both in antagonistic solvents like water and any nonpolar liquid which are considered to represent the extreme cases. Moreover, it is well known that models illustrating such aggregational processes have many applications (for example, binding equilibria in biological applications[168], or in the theory of associated solutions[170] which comprise micellar association as a special phenomenon. This suggests considering suitable models for micellar aggregation phenomena from a slightly more general point of view. In this way it appears easier to obtain a more unifying view with regard to the various models describing self-association processes of soap molecules.

The following treatment is concerned primarily with models which give rise to a "critical micelle concentration" (CMC). It will become apparent in this paragraph, however, that a critical micelle concentration is not sufficient to appraise the formation of micelles in the sense of closed aggregates if there does not exist some aggregate-size-limiting process. Thus taking the case of lipophilic micellar aggregates, a particular concept has to be introduced in order to understand in which way the cooperative hydrogen bond formation leads simultaneously to closed micellar structures.

The general treatment of multiple self-association processes (the most probably encountered physical situation) which describe the formation of a sequence of dimers, trimers etc.[118], is given by

$$
\begin{aligned}
S_1 &\overset{K_1}{=\!=} S_1 \\
2\,S_1 &\overset{K_2}{=\!=} S_2 \\
&\quad\vdots \\
.\,n\,S_1 &\overset{K_n}{=\!=} S_n
\end{aligned}
\tag{1}
$$

where the K_i $(i = 1,2,\ldots n)$ denote the equilibrium constants for the respective equilibria. This yields, considering the mass action law and the conservation of mass in monomer units (s_0) together with the abbreviation $[S_1]/s_0 = \gamma_1$

$$
\sum_{n=1}^{\infty} n\,\gamma_1^n \cdot s_0^{n-1}\,K_n = 1
\tag{2}
$$

which does not predict without special assumptions (see later) a CMC but is an experimentally frequently fulfilled relation for continuous aggregational processes. An alternative scheme would be

$$
\begin{aligned}
S_1 &+ S_1 \overset{K}{=\!=} S_2 \\
S_2 &+ S_1 \overset{K}{=\!=} S_3 \\
&\quad\vdots \\
S_{n-1} &+ S_1 \overset{K}{=\!=} S_n
\end{aligned}
\tag{3}
$$

describing a linear association process where it is often assumed that all the association constants are equal. This case represents a traditionally well-known procedure[130, 170, 168]. Under these circumstances it can be shown by introducing $x = [S_1] K$ and considering Eq. (2) under the condition $K_n = K^{n-1}$ that

$$\sum_{n-1}^{\infty} n \cdot x^n = K \cdot s_0 = \frac{x}{(1-x)^2} \tag{4}$$

from which Fig. 1 is plotted. It is seen that $x < 1$, i.e., $S_1 < K^{-1}$ for any value of s_0 has an upper bound K^{-1} [168]. This particular conclusion depends upon the form of K_n mentioned above. Extensive discussions regarding this case may be found in Prigogine's treatment on Chemical Thermodynamics[170].

The model just mentioned may easily be extended to the case of spherical aggregates and rounded platelets as Poland[168] has shown: introducing, for example, the volume of the aggregate as V_n which is taken to be proportional to n, then the radius is proportional to $n^{1/2}$ and the surface is $\sim n^{2/3}$. Hence, one obtains $K_n = \exp(-an^{2/3} \Delta g_s^{\ominus}/RT) \exp(-bn \Delta g_i^{\ominus}/RT)$ where a and b are geometric factors and $\Delta g_s^{\ominus} = g_s^{\ominus}$ (molecule on the surface) $- g_i^{\ominus}$ (molecule in interior of aggregate) and $\Delta g_i^{\ominus} = g_i^{\ominus}$ (molecule in interior) $- g^{\ominus}$ (molecule in solution).

A micellization, i.e., the transition from an essentially monomeric to an aggregational state, as characterized by the onset of a cooperative aggregation at the critical concentration is thought to take place within a relative narrow weighed-in concen-

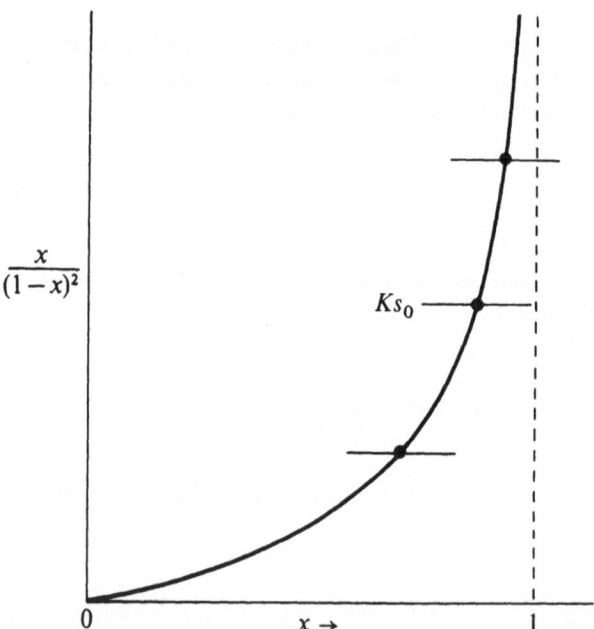

Fig. 1. Schematic representation of the graphical solution of Eq. (4). The horizontal lines indicate various values of Ks_0. The dot at the intersection is the desired solution. $x = 1$ is not allowed solution for finite Ks_0, i.e., x must be less than unity. (Cooperative Equilibria in Physical Chemistry. Oxford: Clarendon 1978)

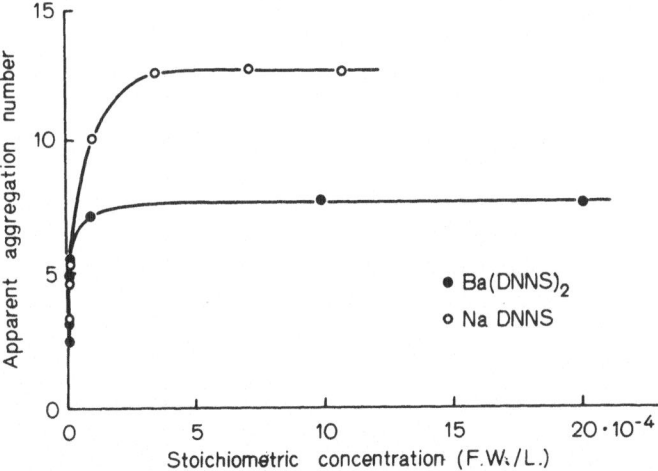

Fig. 2. Concentration dependence of apparent aggregation number of dinonylnaphthalene sulfonates: (○) = NaDNNS, (●) = Ba(DNNS)$_2$ in benzene at 25 °C. Concentration in formula weight (*F. W.*) per liter (*L.*) [J. Colloid Sci. *10*, 139 (1955)]

tration range. The cooperativity of this process is quite essential and warrants a narrow distribution of aggregate sizes centering around an average micellar size. Particularly, with nonpolar detergent solutions one meets many cases of rather narrow apparent micellar weight distributions as may be found, for example, in the work of Kaufman and Singleterry, or Fowkes, especially in[103, 106, 107, 82] and Fig. 2.

2.1.1 Phase — Separation Model

This model represents the most "radical" treatment among the three most frequently discussed approaches to describe micelle formation: it simple postulates ab initio the micellization to be a phase transition. This is justified all the more if large aggregation numbers (like those often encountered in aqueous surfactant solutions) are considered.

Such a phase separation model has been advocated by Shinoda[191, 192], Hutchinson and coworkers[97, 98] and successfully applied, for example, by Fowkes[82] and Singleterry[197] to describe the concentration dependence of the calcium dinonylnaphthalene sulfonate (CaDNNS) and BaDNNS micelles together with the constancy of the monomer activities in the experimentally accessible range of the sulfonate concentrations. This apparent "monodispersity" also observed by other authors[123, 124] is met more frequently with nonpolar solutions of anionic surfactants[65].

In view of this phase concept which is confirmed by the micellization phenomena in many nonpolar detergent solutions, it has been suggested by Eicke and Christen[40] that in line with this reasoning a nucleation step is to be expected (in the approximation of the phase separation model). In order to explain the origin of the energy necessary to overcome the potential barrier associated with the postulated nucleation step, these authors proposed a Monte-Carlo Model of the micellization in nonpolar media[19] using a two-dimensional liquid lattice. With the help of a computer program a phase transition was created which was applied to the formation of micelles. The consideration of the fluctuations in this simulation allows the determination of local variations in density which where directly correlated with the energy and frequency of the nucleation.

The phase separation model follows exactly the description of a two-phase equilibrium, i.e., equating the respective chemical potentials of the particular surfactant in both phases (i.e., monomers in the nonpolar solvent and the micelles) at the critical concentration (CMC). Thus, (assuming ideal condition)

$$\mu_{mic}^{\ominus} = \mu_{solution}^{\ominus} + kT \ln x \, (CMC) \tag{5}$$

which shows that the CMC is a constant at constant temperature. $\mu_{solution}^{\ominus}$ and $\mu_{mic}^{\ominus}$ are the standard chemical potentials of the surfactant molecules in the apolar solution and the "micellar" phases, respectively.

This model which describes a phase transition naturally overemphasizes the cooperativity with respect to the micellization. The surprising monodispersity of various micellar aggregates and the constancy of the monomer activity support the cooperativity concept of the aggregational process. In its simplest form this model does not contain any size limiting step. The latter is principally independent of the cooperativity which had to be included in a consideration of the formation of size limited aggregates. It is thus seen that this model can only be of restricted value towards an understanding of the formation of small particles, usually encountered in nonpolar solutions.

2.1.2 Mass-Action Model

Contrary to the preceding treatment the so-called "mass-action model" develops apparently more naturally from the application of the mass action law applied to the overall aggregation process

$$n S_1 \overset{K_n}{=\!=} S_n \tag{6}$$

where K_n is the association constant of the "all or nothing" process. Together with the conservation of mass (expressed in monomers), one obtains

$$\gamma_1 + n \gamma_1^n \cdot s_0^{n-1} K_n = 1 \tag{7}$$

where s_0 denotes the total weighed-in concentration. The constant K_n can be imagined a product of n−1 individual mass action constants $K_i (i = 2, \ldots n)$, i.e., starting with $S_1 + S_1 \overset{K_2}{=\!=} S_2$. These are thought to be identical according to Eq. (3). Hence,

$$K_n = K^{n-1} . \tag{8}$$

Reducing the number of variables by one in defining $x = K \cdot s_0$ it can be derived from Eq. (6) that

$$\gamma_n = x^{n-1}\gamma_1^n \qquad (9a)$$

which follows from the mass action law. Considering the conservation of mass one obtains

$$\gamma_1 + n\gamma_n = 1 \qquad (9b)$$

where $\gamma_n = [S_n]/s_0$. This is a nonlinear equation with respect to γ_1 which is unsolvable for large n in an analytical form. It is seen from Eq. (9a) that $x < 1$ favors the monomers, $x > 1$ the micellar aggregates and that the transition: monomer → aggregate becomes sharper, the larger the association number n. $x = 1$ corresponds to the transition point to which, accordingly, a critical micelle concentration can be assigned: thus,

$$CMC = \frac{1}{K} = (s_0)_{x=1} \qquad (10)$$

Fig. 3 shows a plot of γ_1 versus log x: the diagram clearly exhibits the shift of the steep slope of γ_1 (monomer contribution) with increasing n towards $x = 1$. It is seen, moreover, that for large n, i.e., $15 < n$, the steepness of the slope stays almost constant, indicating that the cooperativity of the transition: monomer → micelle is rather insensitive with respect to n if n is larger than the above mentioned value.

Generalizing, one could state that the "mass-action" model simulates a cooperative (all or nothing) process with respect to large n values. This model is somewhat

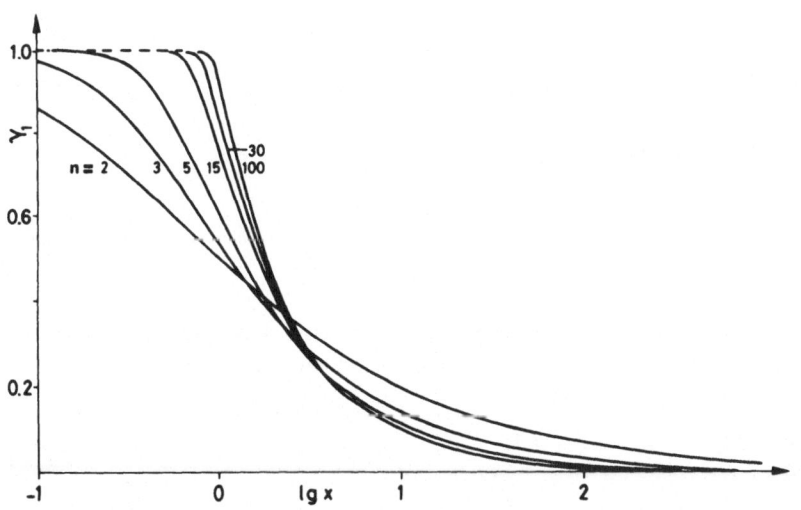

Fig. 3. Monomer fraction γ_1 plotted versus log x, where $x = Ks_0$ and n the aggregation numbers of aggregates

artificial by stressing essentially only two aggregational states; hence it cannot account for a distribution of apparent aggregate molecular weights which is a particular draw-back considering aqueous surfactant systems.

It is to be noted that none of these models considers explicitly the cooperativity by a cooperativity parameter or a size limiting step.

2.1.3 Multiple-Equilibrium Model

According to many experimental observations more or less smooth transitions within a relative small concentration range is observed (such as already described by the mass action model at very low degree of association) with subsequent concentration dependent growth of the aggregates (see e.g.[149]). These aggregates are thought to coexist in mutual equilibrium with each other. Such an association is frequently observed with so-called cationic surfactants (but not at all exclusively) see, for example[53, 105, 123], where dodecylammoniumpropionate in benzene or cyclo-hexane[47, 137] or tridodecylammonium salts in nonpolar solvents[113, 132] may serve as prototypes.

The multiple-equilibrium model which corresponds to the scheme of Eq. (3) does not provide any critical concentration and is considered in the frame of the present discussion to be the opposite limiting case with regard to those detergent systems which had to be assigned to the phase separation model.

It is possible, however, to modify the model in order to account for a cooperativity and a critical concentration: according to the mass action law the concentration of aggregates containing n monomers is [see Eq. (3)]

$$[S_n] = [S_1] \prod_{n=2}^{n-1} K_{n-1}. \tag{11}$$

The conservation of mass with respect to monomers yields

$$\sum_{n=1}^{n} n \cdot \gamma_n = 1 \tag{12}$$

which, if combined with Eq. (11), reads

$$\gamma_1 + \sum_{n=2}^{n} n \cdot \gamma_1^n \cdot s_0^{n-1} \cdot \prod_{n=2}^{n} K_{n-1} = 1 \tag{13}$$

where s_0 is the weighed-in concentration of the surfactant. A cooperative effect is introduced by assuming $K_{n-1} = 0$ for $2 < n < n_0$ and hence one obtains

$$\gamma_1 + \sum_{n=n_0}^{n} n \gamma_1^n \cdot s_0^{n-1} \cdot \prod_{n=n_0}^{n} K_{n-1} = 1. \tag{14}$$

As a special case of Eq. (14) one could consider $K_2 = \sigma K; K_3, K_4, \ldots = K$. By this procedure a cooperativity effect is generated which is controllable by the cooperativity parameter σ[220]. This expression [Eq. (14)] provides a critical concentration and a cooperative feature of the aggregational process. However, above n_0 the growth of aggregates with the concentration is not limited, since no size limiting step has been introduced.

Summarizing the statements of these three most commonly used models, it appears that the so-called "mass action" and "phase-separation" models simulate a third condition which must be fulfilled with respect to the formation of micelles: a size limiting process. The latter is independent of the cooperativity and has to be interpreted by a molecular model. The limitation of the aggregate size in the mass action model is determined by the aggregation number. This is, essentially, the reason that this model has been preferred in the description of micelle forming systems. The multiple equilibrium model as comprised by the Eqs. (10–13) contains no such size limiting features. An improvement in this respect requires a functional relationship between the equilibrium constants and the association number n, i.e., $K_n = f(n)$[154].

In this way the variety of the aggregates is preserved and simultaneously the number of variables (represented by the K_n's) reduced, eventually, to one. Such a procedure is to be discussed in a forthcoming paper by Christen and Eicke[21] and was considered more rudimentarily by Winklmair[220].

2.2 Molecular Interaction Leading to Aggregation or Micellization

The thermodynamic models discussed in the preceeding paragraph provide no insight into the underlying mechanism and molecular interactions leading to aggregational phenomena. The particular value, however, of such models is emphasized by the fact that they apply equally well to both, aqueous and nonpolar surfactant systems.

Regarding the energetic contributions to the free enthalpy of aggregate formation, these are quite different with respect to aqueous and nonpolar surfactant solutions. Moreover, as already mentioned, a survey of experimental results shows that, apart from the nonionic detergents, essentially two classes of ionic surfactants i.e., anionic and cationic have traditionally been distinguished by their aggregational patterns. Accordingly, this difference has also to be interpreted on a molecular basis. Transitions between these groups are frequently observed which reminds one to be aware of the fact that a definition of anionic and cationic detergents is by no means unambiguously possible. It is necessary, therefore, to consider carefully in each case structural and molecular details of the respective surfactant molecule.

Few attempts have been made to explain the formation, stability, association patterns, and properties of surfactant aggregates, in particular, micelles in nonpolar solvents. Principally, the energetic contributions to the total free energy of aggregate formation can be calculated as a chemical potential difference of monomeric detergent molecules due to the transference of the monomers from an (ideal) solution to the interior of the aggregate, i.e., $\mu_{mic}^{\ominus} - \mu_{solution}^{\ominus}$ (see, for example[208]). In detail, the authors proceed from quite different starting points to describe the stability of lipophilic micelles: earlier semiquantitative estimations based upon simple molecular concepts without a clear distinction between enthalpic and entropic contributions were used to determine roughly the total free enthalpy of the aggregate formation. Thus, Pilpel[166, 167] considered three energy contributions, i.e., (i) the change in the interfacial free energy of solute molecules, (ii) the change in their dipole interaction energy, and (iii) the hydrogen bond interaction. It has been shown by the author that the total energy change (ΔE_{tot}) is of the order of 10^{-13} to 10^{-12} erg ($10^{-20} - 10^{-19}$ J) if reasonable values for the parameters determining ΔE_{tot} are used. The result is taken to indicate the existence of sufficiently stable aggregates, i.e., $|\Delta G_{mic}| \gg kT$ (the latter being of the order of $4 \cdot 10^{-21}$ J at 20 °C). However, this estimation is of limited value since various entropic contributions have not been considered which, generally, would counteract the enthalpic part.

A more involved molecular model has been used by Kitahara and Kon-no[120]. It was originally proposed by Flory[75] with respect to the dissolution of polymers. The same idea has been utilized also for compounds with low molecular weights by a number of authors (see[159]). The model describes the transfer of a monomeric surfactant molecule from an assumed ideal solution of detergent monomers to a micelle where the monomer is in a nonideal hydrocarbon environment. The corresponding excess chemical potential of solvent molecules referring to the ideal solution and the outer hydrocarbon portion of the micelle, respectively, is taken to be proportional to the square of the volume fraction of the hydrocarbon portion of the aggregate. Simultaneously, this chemical potential difference can be related to an excess osmotic pressure if diluted solutions are considered[80]. From the equilibrium condition the free enthalpy change of the aggregate formation with respect to the lipophilic (= hydrocarbon) moiety of the surfactants was estimated. The electrostatic interaction between charges of opposite sign (assuming complete dissociation) was taken to be a free energy contribution. This does not appear to be completely correct. Also, all repulsive contributions between like charges have not been taken into account. Thus the total free enthalpy change of the micelle formation represents only an approximate value. Nevertheless, the final equations describe the experimental plot of the aggregation number versus the number of carbon atoms in the hydrocarbon chain satisfactorily if suitable values of the adjustable parameters are chosen. The feature of the model to describe the dependence of aggregation number on the chain length of the detergent moiety would permit statements (considering the above restrictions) as to a suitable selection of surfactants regarding micellization in nonpolar solvents. In addition, the free enthalpy of micelle formation versus the aggregation number yields qualitatively reasonable plots. This is however a feature of all equations, with similar structure describing micelle formation by a superposition of repulsive and attractive terms.

A comparatively more ab initio model with detailed considerations concerning the possible enthalpic and entropic contributions to the free enthalpy of micellization have been carried out by Eicke and Christen[39, 40, 45]. The calculations are based upon the assumption of a particular geometric micellar equilibrium structure, i.e., a cylinder. This structure had been suggested by Peri[161, 162]. The enthalpic contributions were thought to be composed of van der Waals's (solvent-solvent, solvent-hydrocarbon tail of detergent and tail-tail) interactions and of electrostatic interactions within the polar core of the micelle. The latter were, considering the overall effect, repulsive. Assuming reasonable values of the parameters involved in the model a stabilization of the micellar aggregates could be inferred. Additional information was obtained concerning the degree of dissociation of the polar (ionic) portions of the surfactant molecules encaged in the polar micellar core and the solvent dependence of the micellar size. Independent of the before mentioned approach, entropic contributions were considered. Essentially, the entropic loss due to the transfer of surfactant molecules from a nonmicellar to a micellar state have been calculated according to the fundamental Hildebrand[90, 91] concept. This contribution is destabilizing, i.e., it tends to increase the free enthalpy of micellization. The superposition of all the relevant contributions give rise to a $\Delta G_{mic}(\bar{n})$ – plot with a maximum value at small association number $(\bar{n})$[45]. This latter feature has been interpreted as being due to a premicellar aggregate ("nucleus"). The notion is to be understood in line with the concept of a cooperative micelle formation. How far this concept is reasonable will be discussed in connection with kinetic experiments.

Recently, an extensive study regarding a possible prediction of the association patterns of 1, 1-ionic surfactants in solvents with low dielectric constants has been proposed by Muller[151, 152]. His model calculations are based upon a comparison between surfactant association patterns in solution and the formation of ionic lattices by (gaseous) alkali halides. The application of this latter process to the aggregation of detergent molecules in solution is certainly questionable with regard to many details, particularly, the considerations of internal rotational and vibrational entropic contributions in evaluating the stability of molecular clusters in solution. It might be conceded that some loose similarities exist between initial association steps of ionic surfactants in solution and the building up of a solid state ionic lattice. Muller's model predicts, essentially, a stepwise sequential formation of open chain oligomers with approximately equal equilibrium constants for the binding of additional monomers if the sum of the radii of the ionic headgroups of one surfactant molecule is large. If, on the other hand, this sum is small, "compact" clusters are preferred corresponding to this model. At first sight the predictions of this model do not appear unreasonable in view of the available experimental material. The value of the treatment is seen in the fact that it attempts to overcome the artificial distinction between cationic and anionic surfactants.

However, apart from the generally known problem of selecting suitable parameter values which necessarily weaken any detailed analyses, new experimental information and insight into the formation of lipophilic micelles has to be discussed in the following treatment which essentially supersedes the conclusions of the above model: recent photon correlation experiments with the system $H_2O/AOT/i-C_8H_{18}$, pertinent IR and NMR investigations of lithium and cesium salts of dinonylnaphthalene sulfonic acid in heptane[201] and of alkali and alkylammonium polystyrene sulfonates[228] as well as vapour-pressure osmometric measurements with alkali and alkylated quaternary ammonium di-2-ethylhexylsulfosuccinates[46] strongly suggest hydrogen bonding to be of predominant, if not decisive, importance concerning micellization in nonpolar media.

In this connection the surprising thermal stability of AOT micelles in nonpolar solvents has been demonstrated again with the help of photon correlation spectroscopic experiments[227] in the AOT/isooctane system between −85 and +95 °C. This was explained by the formation of a hydrogen bond network where it was assumed that $\Delta H_{binding}$ for each subsequent hydrogen bridge is constant. The stability of the micellar aggregate is then determined by the total number of hydrogen bridges times $\Delta H_{binding}$ of a single bond.

These considerations are confirmed, in particular, by Oedberg et al. who concluded from their IR investigations with Li- and CsDNNS micelles in heptane that all the water molecules in the reversed micelles are hydrogen bonded. It appears worth mentioning that the first water molecules added to the system induced deviations from the Lambert-Beer law for LiDNNS and large intensity changes of the vibrations of the sulfonic acid group in CsDNNS.

Corresponding results were derived by these authors from NMR measurements. The variations of the chemical shift towards higher fields is rather small indicating that already the first water molecules participate in a hydrogen bonded structure (see also Zundel). Figure 4 shows that the water resonance is broadened at low w_0-values (= $[H_2O]/[surfactant]$). This finding can be interpreted to be due to a decreased mobility of the first water molecules taken up by the micelle and, thus, to be a consequence of hydrogen bonding to the sulfonic acid groups. Also, it turned out that the NMR spectral lines were broad at low w_0-values with respect to LiDNNS and sharpened with increasing degree of hydration. Considering CsDNNS, the spectrum exhibited sharp lines even with dried samples. These results indicate, as do the IR data, more restricted motions of the naphthalene skeleton in LiDNNS compared to CsDNNS. Quite probably, the strongly interacting lithium ions (via the hydrogen bridges of the water of hydration) promote a denser micellar structure. This conclusion agrees nicely with Zundel's observations concerning the decreasing strength of the hydration interaction in proceeding from the lithium to the cesium salts. Moreover, these results are in perfect agreement with conclusions obtained by Ekwall and coworkers[64] in their comprehensive work on binary and ternary AOT systems.

The above reported work is supplemented by the fact[127] that in the presence of water the so-called effective ionic radius (as defined by considering the hydration sphere) of the ammonium is according to Conway[25] 2.5 Å. This is comparable to that of cesium or rubidium. With regard to sterical considerations, it is par-

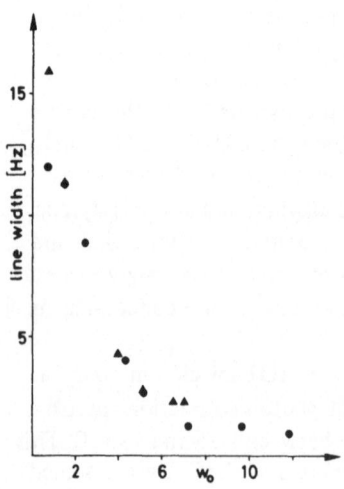

Fig. 4. Line width of water NMR resonance lines in LiDNNS (•) and CsDNNS (▲) micelles in heptane at 23 °C and 20 °C, resp. plotted versus $w_0 = [H_2O]/[surfactant]$ (In: Ion Exchange and Membranes Vol. 2, 83 (1975). Gordon and Breach)

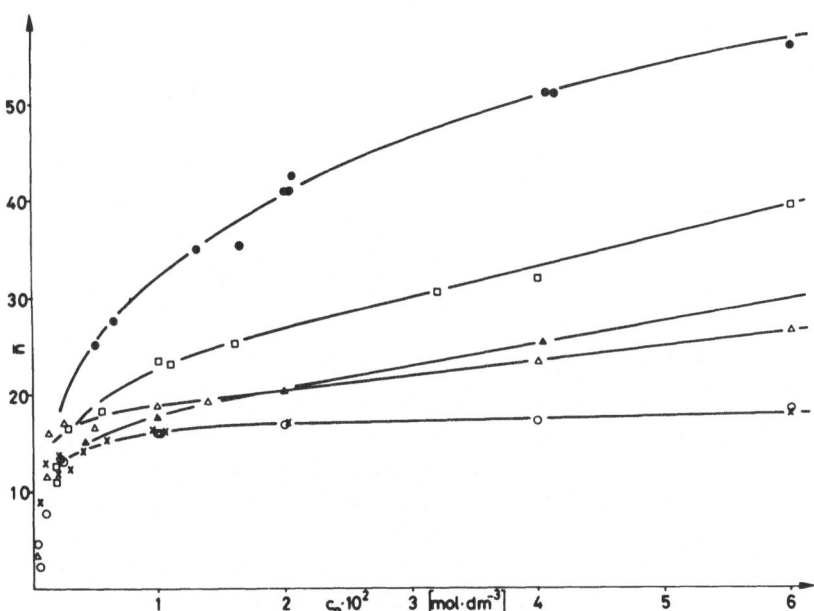

Fig. 5. Average aggregation numbers ($\bar{n}$) of Li-(x), Na-(o), K-($\triangle$), NH$_4$-($\blacktriangle$), Rb-($\square$) and Cs-($\bullet$) di-2-ethylhexylsulfosuccinate in isooctane at 25 °C versus weighed-in concentration. (VPO measurements)

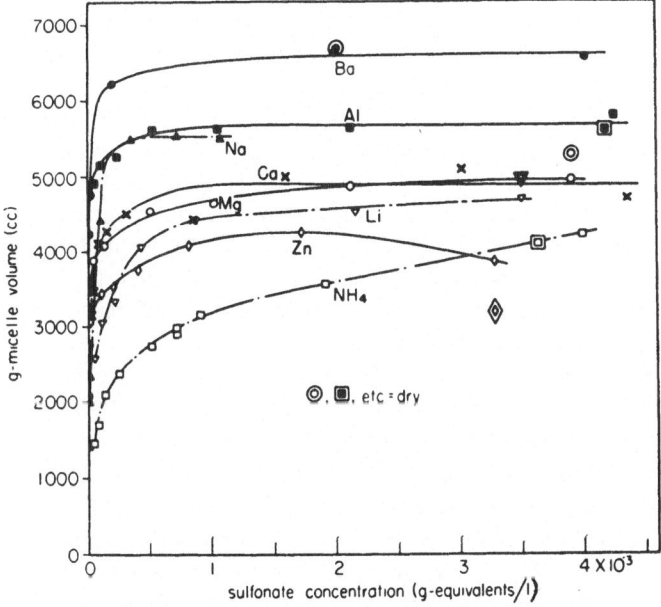

Fig. 6. g-micelle volumes ($\sim$ aggregation numbers = acid residues per aggregate) in ml as determined by fluorescence depolarization measurements versus sulfonate concentration in g-equivalents per liter of dinonylnaphthalene sulfonates at 25 °C in benzene saturated with water [J. Colloid Sci. *12*, 465 (1957)]

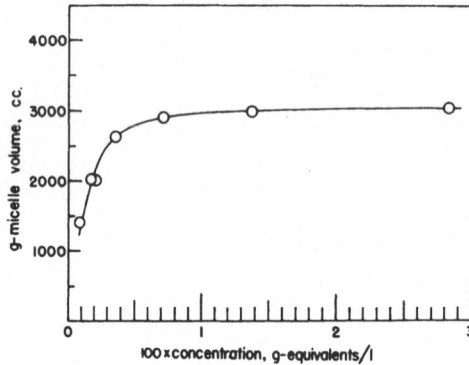

Fig. 7. g-micelle volumes ($\sim$ aggregation numbers = acid residues per aggregate) of lead 2-ethylhexylsebacate by fluorescence depolarization measurements at 26 °C versus weighed in concentration of soap in g-equivalents per liter in benzene saturated with water [J. Phys. Chem. *62*, 1257 (1958)]

ticularly interesting that the distance of closest approach of another ion (for example, a di-2-ethylhexylsulfosuccinate ion) is identical both for the sodium and the tetramethylammonium ion[127]. Moreover, it turns out that replacing the sodium by the ammonium ion the well-known pattern of an almost concentration independent aggregation number of AOT is lost and instead a pronounced concentration dependence is exhibited, see Fig. 5. The diagram also shows the different alkali salts derived from AOT which demonstrate that, except for the sodium and lithium salts none of the other alkali derivatives of di-2-ethylhexylsulfosuccinates displays a concentration independent aggregation number in isooctane. Exactly the same behavior has been observed by Kaufman and Singleterry[105, 107] with the analogous dinonylnaphthalene sulfonate detergents (see Fig. 6 and 7)[1]. If the sodium is replaced by the alkylated quaternary ammonium ions, then in addition to the concentration dependent aggregation number, the degree of association (= average size of the particles) decreases to three or four monomers at about 10^{-2} mol dm^{-3} weighed-in surfactant concentration (Fig. 8).

These observations indicate a decreased hydration interaction between water and the ammonium ion (or its derivatives) as compared to the sodium ion. The findings are confirmed by unpublished data[131] regarding AOT micelles in isooctane saturated with D_2O. The smallest aggregates (micelles) observed with the photon correlation technique are smaller which would follow from the fact[67] that the heats of ionic hydration are smaller and the hydrogen bonds (OD . . . X) to a foreign acceptor (X) are weaker in D_2O compared to H_2O.

All these facts are in agreement with results obtained by Zundel and coworkers[228] from IR investigations on 5 μ thick membranes of polystyrene sulfonates under conditions of controlled humidity and temperature. From these experiments with, for example, sodiumpolystyrene sulfonate membranes at low degree of hydration it was concluded that one water molecule is attached to the counterion (= sodium) which connects via two hydrogen bridges simultaneously two other sulfonate molecules, thus forming a trimer (Fig. 9). It is known that such trimeric subunits were frequently detected in the case of AOT[38, 40, 43]. Zundel observed, in particu-

---

1 The explanation provided by these authors to interpret the association patterns of the lead 2-ethylhexyl sebacate in benzene has to be modified according to the present view: sodium and lead are quite comparable with respect to their hydration interaction[25].

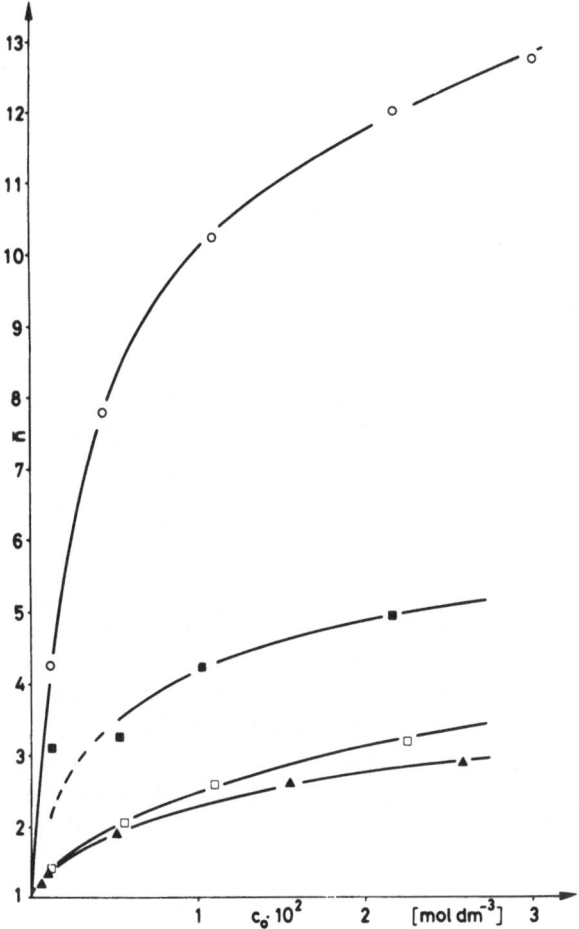

Fig. 8. Average aggregation number ($\bar{n}$) determined by VPO measurements versus weighed-in concentration of surfactant of NH_4- (○), $N(CH_3)_4$- (■), $N(C_2H_5)_4$- (□) and $N(C_3H_7)_4$- (▲) di-2-ethylhexylsulfosuccinates in benzene at 25 °C. [Helv. Chim. Acta 61, 2258 (1978)]

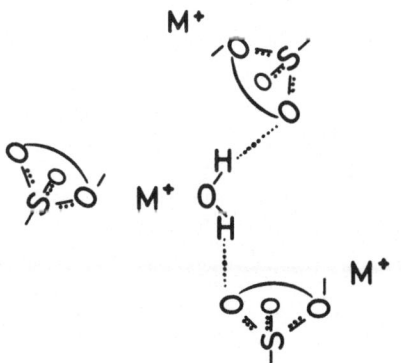

Fig. 9. Formation of trimeric aggregates of 1,1-ionic surfactants with sulfonate groups. [Hydration and intermolecular interaction. New York: Academic 1969]

lar, the OH-stretching vibrations of such hydrogen bridges originating from water molecules which show a decreasing interaction in proceeding from the sodium to the ammonium ions. This interaction is especially weak with cations like $N(CH_3)_4^+$, $N(C_2H_5)_4^+$ etc. The OH-stretching vibrations are shifted in the latter case closely towards the wave numbers corresponding to those found in pure bulk water.

Considering the above reported experimental material, it now appears to be certain that the concept of hydration interactions with following hydrogen bond formation within the reversed surfactant aggregates explains most satisfactorily all experimental details from a unified point of view.

The preceding discussion has to be judged with the view that minute amounts of water in all organic systems, even under particular precautions, cannot be completely excluded. It has been shown[53] that the amount necessary to form a minimum number of hydrogen bridges sufficient to support growing of the aggregates and to stabilize the final particles is far below the detectable amount of water in organic solvents.

These statements lead to some interesting consequences: the question arises wether or not a surfactant in oil can be regarded as a binary system. It should be kept in mind that frequently the so-called critical micelle concentration (CMC) is observed at weighed-in concentrations as low as $10^{-5} - 10^{-4}$ mol dm^{-3} or even lower (see, e.g.[104]). A water to surfactant ratio of 1:1 represents a very small, hardly detectable amount of water. But, as has been pointed out, the necessary water concentration can be by far smaller and still support the growing of aggregates. It should be pointed out that water and surfactant are present in nonpolar solution in comparable amounts. The system could, therefore, be considered as being ternary. If this is anticipated, then the "CMC" would loose its significance with respect to a surfactant/solvent property since it is changing with the water content[53]. Only if extremely dried surfactant is used then the CMC appears to be almost independent of the water content of the solvent (see Fig. 10).

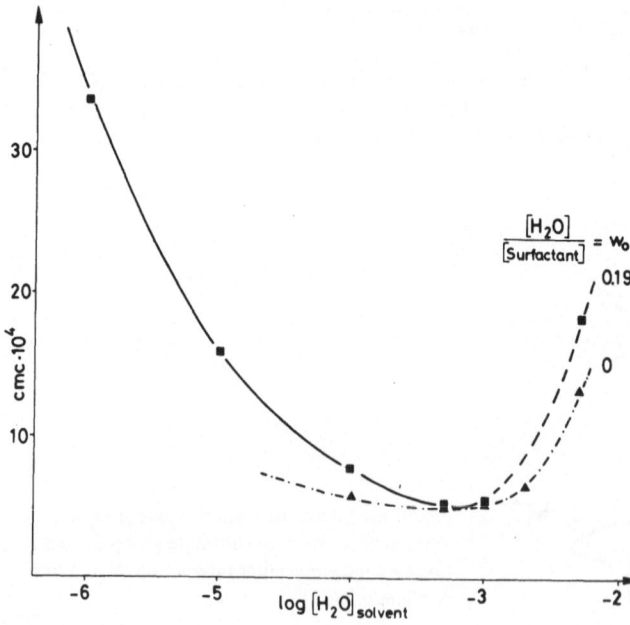

Fig. 10. CMC versus water content of apolar solvent. Parameter: $[H_2O]/[\text{surfactant}]$ = water initially attached to the surfactant. [Helv. Chim. Acta *61*, 2258 (1978)]

3 Micellar Size and Shape in Nonpolar Surfactant Solutions

3.1 Dependence on Molecular Structure

Contrary to aqueous surfactant systems of comparable surfactant concentrations, i.e., $10^{-3} < c_0 < 10^{-1}$ mol dm^{-3}, where micelles are generally found to be of spherical shape due to the repulsive interaction of the ionic charges in the micelle/solution interface and long-range electrical forces between the charged aggregates, there does not exist such natural restriction of reversed micellar aggregates with respect to deviations from a spherical shape. Thus, apart from considerations with regard to the energies responsible for the formation of colloidal aggregates and their stability in nonpolar media, size and shape of these entities and their dependence on the molecular structure and environmental conditions deserve more closer examination.

3.1.1 Dependence on the Hydrocarbon Moiety

It appears to be generally accepted that sterical restrictions are an important factor regarding the formation of reversed micelles, i.e., the relative bulkiness or (more precisely) the ratio of the cross-sectional areas of the hydrocarbon to the polar portions essentially affects the size and shape of the aggregates[162]. The cross-sectional area of the hydrocarbon moiety of most oilsoluble surfactants exceeds that of the polar portion (headgroup). For such surfactant molecules the spherical micelle allows the loosest packing of tails around a core of a given volume, and the prolate ellipsoid the next with regard to packing density, while an oblate ellipsoid of equal eccentricity requires the closest packing[197]. The prolate ellipsoid is then a more probable form than the oblate ellipsoid[213] as the micelle departs from a spherical shape. The dinonylnaphthalenesulfonates[82, 105] represent examples where the growth of micelles beyond spherically shaped aggregates appears energetically to be unfavorable. A similar case is mentioned by Heilweil[89] who investigated sodium 2,6-di-n-octyl- and sodium 2,6-di-n-dodecylnaphthalene sulfonates in n-decane, n-heptane and benzene. In n-decane the apparent micellar aggregation numbers were essentially independent of C_8- and C_{12}-chain lengths.

Deviations from a spherical shape towards a prolate ellipsoid (cylinder) has been reported experimentally in the case of AOT[64, 162]. A lamella structure should be even less probable, although it might occur if the cross sections of the headgroup and the tail were equal, as in normal alkali fatty acid soaps. The lamellae may grow in the latter case into crystals too large to be considered micelles. Honig and Singleterry[93] investigated phenyl stereates, molecules having a relatively slim tail. These may be packed in such a way that growth of the micelle in one or two directions becomes possible[175]. In this case, since there is no specific preference for a particular micellar size, the size distribution is expected to be broad.

Debye and coworkers[31, 32] observed a decrease in micellar size with increasing chain length in the case of α-monoglycerides with varying carbon-numbers from C_{10} to C_{18} in benzene and chlorobenzene using light-scattering and vapor-pressure osmometry. From IR measurements these authors inferred that intermolecular H-bonding plays a role in the association of monoglycerides (see also the temperature dependence). From the large difference, however, between weight- and number average molecular weights they concluded that probably no micelle formation takes place.

Sirianni and coworkers[198] studied again α-monostearin in benzene by means of vapour pressure osmometry and concluded that the solution consists of only monomers and dimers. Nevertheless, these authors reported a CMC which was in satisfactory agreement with the value found by Debye and Prins[32]. On the other hand Robinson[178] reported light-scattering data regarding the same system which conformed excellently to Debye's results. This apparently contractictory behaviour should be discussed with respect to a paper by Becher[10] who points out that Debye's conclusion inferred from the large difference between number and weight-average molecular weights with respect to the formation of micelles is unwarranted'due to the considerable difference in the concentration dependence of the apparent number and weight-average aggregation numbers (Fig. 11). These comments refer especially to cationic and nonionic surfactants exhibiting relative large CMC-values. In this case the high monomer concentration may considerably falsify the number average molecular weight of the aggregates. An interesting example of a sterical effect regarding a possible micelle formation of nonionic surfactants has been reported by Becher[9]. The data refer to monostereate esters of the dianhydrohexitols. The 1,4:3,6-dianhydrides of the isomeric hexitols D-glucitol (sorbitol) D-mannitol, and L-iditol consist of two fused tetrahydrofuran rings, to each of which a hydroxyl group was attached. The rings were tilted towards each other. The hydroxyl group may be oriented, accordingly, in one of two ways: it can be oriented into the fold between the

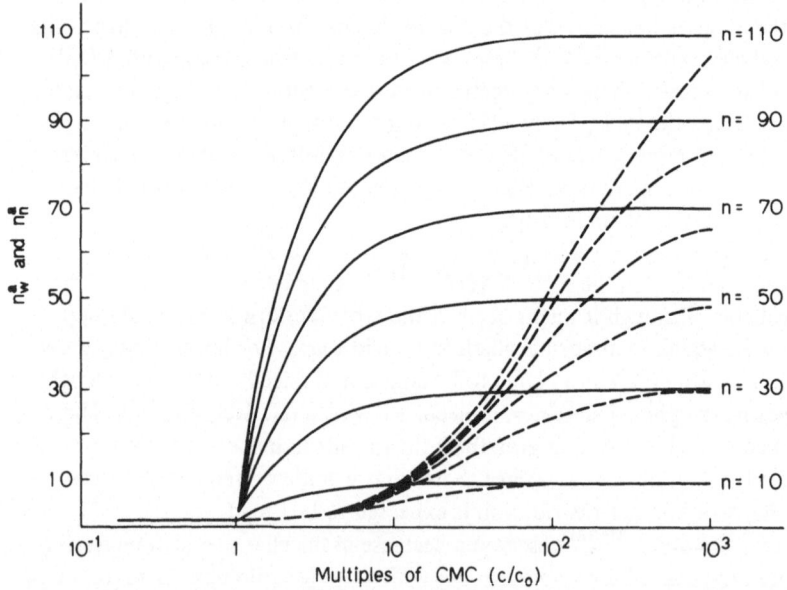

Fig. 11. Apparent weight-average (n_w^a solid lines) and number average (n_n^a, dashed lines) aggregation numbers as function of log (concentration) for a number of n-values. (The concentration has been converted into multiples of the CMC) [Nature *206*, 611 (1965)]

two endo-hydroxyl rings, or away from the fold (= exo-hydroxyl). The results from light-scattering measurements showed that only the esters of the endo-hydroxyl form micelles. This micelle formation appeared to be independent of the orientation of the free hydroxyl. These results were explained by Becher qualitatively in terms of the steric possibility of the chain of the exo-ester orienting itself in such a way as to shield the hydrophilic moiety, thus rendering micellization unnecessary.

Fendler and coworkers[70] applied H^1-NMR studies to investigate the effect of carbon chain length variations on the aggregate size. These authors selected octyl-ammonium carboxylates in benzene and carbon tetrachloride, i.e., propionate, buty-rate, and tetradecanoate. They observed a strong CMC dependence on the C-chain length with opposite effects, depending on whether the C-chain is attached to the ammonium ion or to the carboxylate groups. However, this finding refers only to benzene solutions. In carbon tetrachloride no such dependence could be found, in-dicating the particular role of the solvent in nonpolar surfactant systems.

In another paper Fendler et al.[59] investigated again the CMC dependence of alkylammonium propionates on the number of carbon atoms in the alkyl group of the quaternary ammonium cation (Fig. 12). The same feature as mentioned with octylammonium carboxylates is observed, i.e., the constancy of the CMC values for these alkylammonium propionates in carbon tetrachloride, while in benzene a linear plot is obtained. The latter is taken to indicate the decreased solubility of the mono-mers with increasing alkyl chain length in benzene. Table 4[71] exhibits in addition

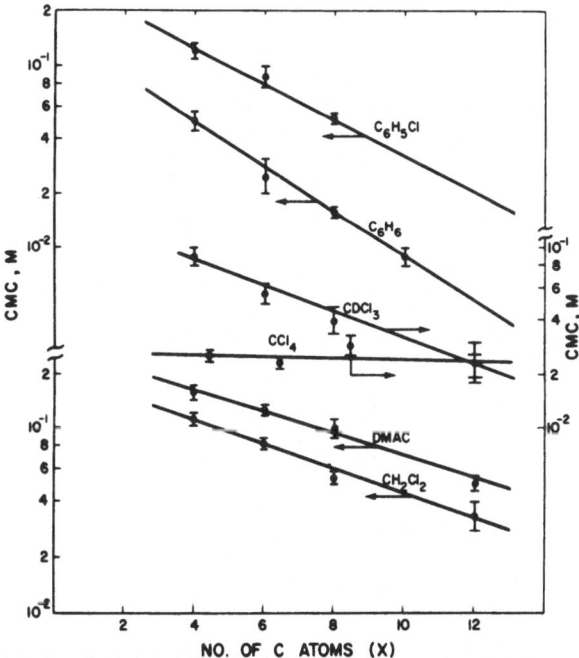

Fig. 12. CMC of alkylammoniumpropionates in $CDCl_3$, CH_2Cl_2, C_6H_5Cl and DMAC as a func-tion of the number C-atoms (x) in the alkyl chain of BAP (butyl-), HAP (hexyl-), OAP (octyl-), DAP (dodecylammoniumpropionate) [J. Phys. Chem. 77, 1876 (1973)]

Table 4. Micellar parameters in benzene and carbontetrachloride. BAP (butyl-), HAP (hexyl-), OAP (octyl-), DeAP (decyl-), DAP (dodecyl-) ammonium propionates. [Faraday Trans. I, *68*, 280 (1973)]

	C_6H_6			
Surfactant	CMC obs./M	CMC calc./M^b	n	K/M^{1-n}
BAP	$(4.5-5.5) \times 10^{-2}$	2.9×10^{-2}	4	1×10^4
HAP	$(2.2-3.2) \times 10^{-2}$	5.5×10^{-3}	7	5×10^{12}
OAP	$(1.5-1.7) \times 10^{-2}$	6.7×10^{-3}	5	1×10^8
DeAP	$(8-10) \times 10^{-3}$			
DAP	$(3-7) \times 10^{-3}$			

	CCl_4			
Surfactant	CMC obs./M	CMC calc./M^b	n	K/M^{1-n}
BAP	$(2.3-2.6) \times 10^{-2}$	1.9×10^{-2}	3	9×10^2
HAP	$(2.1-2.4) \times 10^{-2}$	7.8×10^{-3}	7	7×10^{11}
OAP	$(2.6-3.1) \times 10^{-2}$	8.0×10^{-3}	5	5×10^7
DeAP	$(2.2-2.7) \times 10^{-2}$	1.1×10^{-2}	5	1×10^7
DAP	$(2.1-2.5) \times 10^{-2}$	1.7×10^{-2}	4	5×10^4

a obtained from treatment of $C\underline{H}_3C\underline{H}_2CO_2^-$ chemical shifts at 33 °C;
b using eqn (10)[71]; see Discussion[71].

the interesting fact that the aggregate sizes are apparently independent of the solvents used in this investigation. The aggregation numbers are small. It is, however, questionable whether an aggregate composed of three monomers should be termed

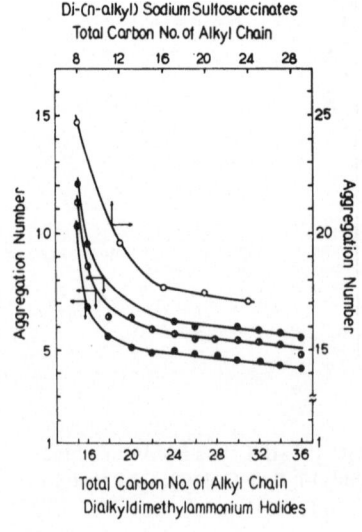

Fig. 13. Aggregation numbers as function of number of carbon atoms in alkylchains of di- (n-alkyl) sodium-sulfosuccinates and dialkyldimethylammonium halides. [J. Colloid interface Sci. *35*, 636 (1971)]

"micelle" and whether the notion "CMC" is appropriate. The equilibrium constants are again different in both solvents in agreement with the above discussed CMC-dependence.

Finally, Kon-no and Kitahara[123] followed the chain length dependence of the aggregate sizes with di(n-alkyl)-sodium sulfo succinates and dialkyldimethylammonium halides, see Fig. 13.

3.1.2 Dependence on the Counterions

Not only sterical effects due to the molecular structure of the surfactant tails have to be considered but also the counterions seem to affect the size and shape of the reversed micellar aggregates. Such effects have been observed, for example, by van der Waarden[213] who found with alkalinaphthasulfonates in heptane and toluene an increase of the aggregates by varying the cations in the order H^+, Li^+, Na^+ and K^+, or Mg^{2+}, Ca^{2+}, and Ba^{2+}, i.e., in the order of decreasing hydration interaction. Similar findings were reported by Reering[176] with sodium and potassium-(tripentylmethyl)benzene sulfonates (TPMBS) in heptane and by Kitahara[123] with sodium and potassium di-2-ethylhexylsulfosuccinates in benzene, carbontetrachloride, and cyclohexane. Finally, Eicke and Christen[39] repeated these investigations with all alkali derivatives of Aerosol OT (alkali di-2-ethylhexylsulfosuccinates) in different solvents (benzene, cyclohexane, carbontetrachloride, n-pentane and C_4H_4S). These measurements have been reproduced again because of their intrinsic importance regarding micelle formation in nonpolar media by Eicke and Hammerich (unpublished results, see Fig. 5). With increasing effective ionic radii of the alkali ions there is a decreasing tendency to form concentration independent aggregates. The aggregates are, however, considerably increasing in the average association number with a size distribution becoming broader with decreasing hydration interaction.

Furthermore, Reerink, found a remarkable concentration dependence of KTPMBS aggregates with a broad size distribution which would be in accord with the assumption of equal equilibrium constants for each successive association step. This behaviour sharply contrasted with the findings regarding the sodium TPMBS micelles where the author derived a sharp size distribution of the aggregates from the satisfactory coincidence between number and weight-average micellar weights. The aggregate shape appeared to be best represented by short rods[170].

Kitahara's investigations using sodium and potassium di-2-ethylhexylsulfosuccinates resulted in an increase in the apparent micellar weight by replacing the sodium by potassium, which is confirmed by the measurements of Eicke and Hammerich. Quite surprisingly, however, the corresponding phosphate compounds, i.e., the alkali di-2-ethylhexylphosphates, seem to exhibit a decrease in the average aggregate sizes in the order from sodium to cesium[39]. If these experiments prove to be correct an explanation might be found in the rather inflexible hydrocarbon tails due to the structure of the phosphate compound which fail to shield sufficiently the polar groups. Along the lines of the preceding discussion regarding the importance of the hydrogen-bridges for the formation and stabilization of reversed micelles the findings with alkali TPMBS and alkali di-2-ethylhexylsulfosuccinates are in close agreement with the above discussion. The effect of counterions on the CMC of cationic surfactants in nonpolar media has been studied by Muto et al.[154] with the help of TCNO solubilization.

Recalling Zundel's[228] investigations it is now certain that the hydration interaction of the alkali ions decreases in the order from the sodium to the cesium ion. This, then, reduces the stability and causes concentration dependent aggregates. The magnitude of the destabilization will depend on the respective effects of the counterions and the nature of the hydrocarbon tails. It is readily visualized, therefore, that in the case of the alkali tripentylmethylbenzene sulfonates an unfavorable balance of the above mentioned effects leads, already for the potassium ion, to a considerable destabilization of the aggregates. According to Reerink[176] only the relative large loss in the standard entropy ($\Delta S^{\ominus} = -15$ cal degree^{-1} mol^{-1}) due to the micellization prevents the formation of excessively large aggregates.

Sterical effects of the effective (or crystallographic) ionic sizes in the case of the alkali di-2-ethylhexylsulfosuccinates as supposed by Kon-no and Kitahara[123] to explain the variation in the association patterns is, according to the above discussion, of only secondary importance. This all the more, since the size of the anion is certainly larger, even compared to cesium, thus exceeding by far the space requirements of any alkali cation.

3.2 Dependence on Environmental Conditions

3.2.1 Solvents

It had soon been recognized that the stability of reversed micelles is considerably solvent dependent (see[139], Table 3). The tendency to form micelles decreases in general with increasing polarity of the solvent. However, rather "weak" polar solvents like chloroform or the pseudo-apolar solvent dioxane in which micellization was to be expected according to the low dielectric constant, did not fit into this too simplified concept. In particular the frequently reported contradictory observations regarding (apparent) similar solvents (as viewed from comparable dielectric constants) as, for example, cyclohexane, carbontetrachloride, or benzene can be explained by strong solvation interactions between the contact ion pairs of the surfactant molecules and the π-electrons of the benzene or between the surfactant dipole and the highly polarizable carbontetrachloride molecule (see e.g.[18, 112]). The remarkable solvating tendency of benzene has recently been demonstrated by Eicke and Denss[52]: these authors observed a considerable decrease of the catalytic activity of dodecylammoniumpropionate in cyclohexane already at small mole fractions of benzene. Simultaneously it turned out that the solvation interaction by benzene competes strongly with the hydration interaction regarding the ammonium ion. Therefore, a considerable number of detailed analyses were carried out concerning size and shape dependences of the aggregates on the solvents used. Moreover, it was thought to obtain information on intramicellar interactions, for example, on the tendency to form hydrogen bonds.

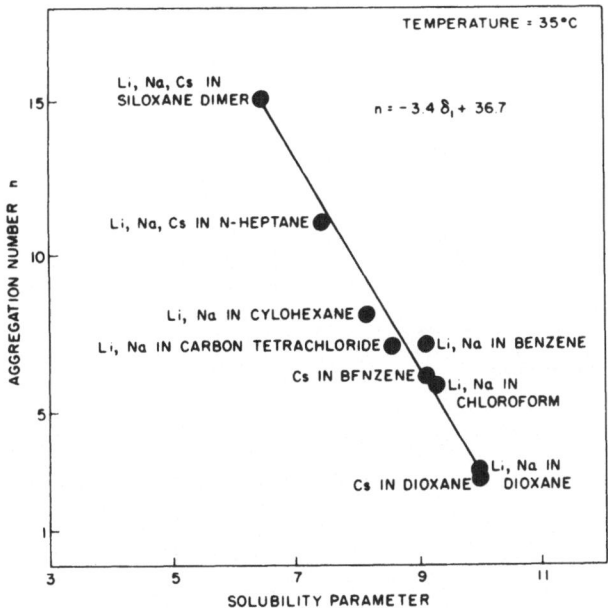

Fig. 14. Aggregation number (n) of sodium dinonylnaphthalene sulfonate versus solubility parameter of the solvents. [J. Phys. Chem. *68*, 3453 (1964)]

A comprehensive paper by Little and Singleterry[134] was devoted to the solubility of alkali dinonylnaphthalene sulfonates in different solvents with particular
emphasis on the mutual relation between the solubility parameter and micellar size,
see Fig. 14.

The solubility parameter introduced by Hildebrand[90], rather than the dielectric constant or dipole moment is a characteristic quantity of the solvent which appears appropriate (if no specific solvation effects have to be taken into account) to
forecast the micellar solubility of the alkali dinonylnaphthalene sulfonates in the
particular solvent. As the solubility parameter of the solvent is increased, the micelles
tend to assume a smaller size (Fig. 14). This size reduction gives a looser packing
of the DNNS tails and, thus, exposes the more interactive aromatic and polar parts
in such a way as to reduce the difference between the solubility parameter of the
solvent and the effective solubility parameter of the solvent-accessible portions of
the lipophilic micelle. The automatic matching of the solubility parameter for micelle and solvent by reduction of micelle size and packing in solvents of high solubility
parameters recalls the behavior of linear macromolecules in solvents of different solvent power.

In particular, Fig. 14 shows that the sodium dinonylnaphthalenesulfonate aggregates decrease in size approximately linearly with increasing solubility parameters
between 6.5 and 10. According to Little[135] the barium dinonylnaphthalene sulfonate (which forms spherical micelles independent of concentration) also obeys
a linear relation between micelle size and the solubility parameters (see Fig. 15).
Actually, quite a number of authors recommend this relationship, for
example[117, 118, 124, 142] as the most reliable and appropriate one. It should be noted,

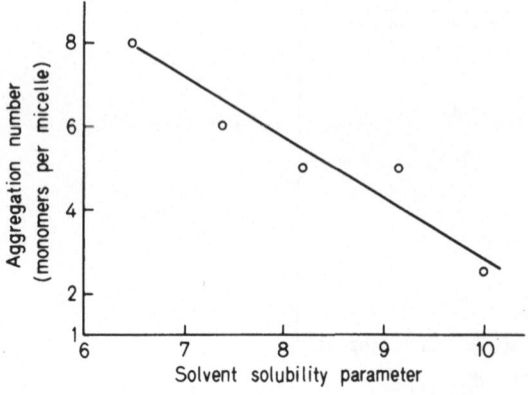

Fig. 15. Aggregation numbers of barium dinonylnaphtalene sulfonate versus solubility parameter of solvents. [J. Phys. Chem. *74*, 1817 (1970)]

however, that there exist some reports about failure of such a relationship (compare, e.g.[7]).

Such deviations may be traced back in some cases to a considerable distortion of the solvent structure by the solute molecule, a well-known effect, for example, with molecules possessing large dipole moments (see, e.g.[84]).

A considerable solvent effect has been reported by Heilweil[89] for sodium 2,6-di-n-dodecylnaphthalene-1-sulfonates in benzene and n-decane. The author finds an apparent molecular weight in benzene of 5.500, about half the value in n-decane. More recently, Eicke and Christen[39] investigated again the solvent effect on the micellar size by plotting a correlation diagram of the average aggregation number

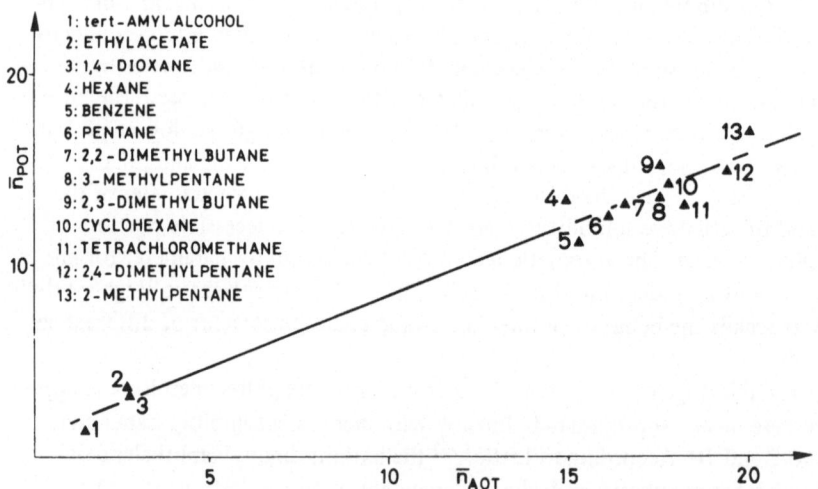

Fig. 16. Correlation diagram of the mean aggregation number per micelle of AOT ($\bar{n}_{AOT}$) and sodium di-2-ethylhexylphosphate ($\bar{n}_{POT}$) in different solvents, 25 °C. [J. Colloid Interface Sci. *46*, 417 (1974) (corrected according to new results)]

for sodium di-2-ethylhexylsulfosuccinate and sodium di-2-ethylhexylphosphate in thirteen different solvents (Fig. 16). The Figure characteristically reflects the nature of the solvents in the aggregation number of the surfactants and the predominant effect of the tails which are almost identical for both surfactants. Peri[162] has reported larger deviations from a straight line than that given in the above diagram (Fig. 14). This could be due to impurities of the AOT samples as mentioned by the author.

Recent results derived from the positron annihilation technique introduced by Ache and coworkers[101] into the study of aqueous and nonpolar detergent solutions are particularly interesting (Fig. 17). Dodecylammoniumpropionate (DAP) has been investigated in benzene, cyclohexane and n-hexane as well as AOT in benzene. The intensity I_2 of the longlived (ortho) positronium (o-Ps) was detected. The latter originates from reactions and subsequent annihilation of thermalized or nearly thermalized o-Ps atoms (see "Methods"). The drastic changes at a particular detergent concentration is remarkable. The breaks in I_2 versus surfactant weighed-in concentration were assigned to the so-called CMC of the respective surfactant-solvent system. Such an assignment is not unambiguous and cannot be done properly without additional information regarding the investigated system. It appears, therefore, advisable to speak more generally of an apparent critical concentration, especially in the case of DAP (see the preceding paragraph). The shifts of these critical concentrations with changing the solvent show the expected trend, i.e., the more nonpolar the solvent, the lower the critical concentration due to a decreasing amount of nonassociated sur-

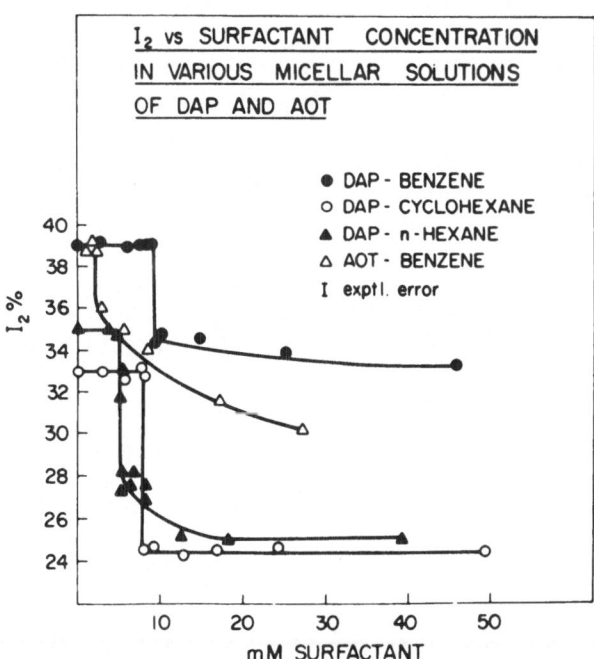

Fig. 17. Intensity (I_2) of the longlived thermalized orthopositronium versus surfactant concentration in various micellar solutions of DAP and AOT, 20 °C. [J. Am. Chem. Soc. *100*, 984 (1978)]

Table 5. CMC values as obtained by Positron annihilation techniques and literature data. [J. Am. Chem. Soc. *100*, 984 (1978)]

Values[a]		CMC, mM	
Surfactant	Solvent	This study	Lit.
DAP	Benzene	8.75 ± 0.25	3–7
	Cyclohexane	8.26 ± 0.25	8[b]
	n-Hexane	6.15 ± 0.45	6[c]
AOT	Benzene	2.20 ± 0.10	2.0–2.7
NaDS	Water	32±1	33
DTAB	Water	57±1	65

[a] Reference 2b.
[b] Dodecylammonium butanoate.
[c] In octane.

factant molecules. Table 5 shows a comparison of the critical concentrations thus determined with corresponding data from the literature. According to the satisfactory agreement it appears that the positron annihilation technique may actually supply conventionally defined CMC values.

A considerable solvent dependence has been reported by Debye and coworkers[32, 33] in their already mentioned study of α-monoglycerides in benzene, chlorobenzene, and chloroform. The authors found a correlation between the cohesive energy densities of the solvents and the clustering tendency of the nonionic surfactants: with decreasing cohesive energy the aggregation increased. In chloroform no micelles were found.

Kon-no and Kitahara[124] made a comparison of anionic, cationic and nonionic surfactants regarding their association in nonpolar solvents. These authors chose poly-

Table 6. Comparison of aggregation numbers among anionic, cationic, and non-ionic surfactants and solvent dependence of aggregation numbers: $NaD2EC_6S$ = AOT, $D2EC_6ABr$ = Di-2-ethylhexylammonium bromide, $DC_8P(EO_{9.2})$ OH = polyoxyethylene 2(1,3-dioctoxypropyl)ether [J. Colloid Interface Sci. *35*, 636 (1971)]

Solvent	δ^a	$NaD2EC_6S$	$D2EC_6ABr$	$DC_8P(EO)_{9.2}H$
Benzene	9.15	13.6	2.3	1.0
Carbontetrachloride	8.6	20.6	2.2	1.0
Cyclohexane	8.2	18.2	4.0	1.0
n-Heptane	7.45	21	4.7	1.0
Isooctane	6.85	21	4.7	1.0

[a] δ: Solubility parameter of solvents at 25 °C.

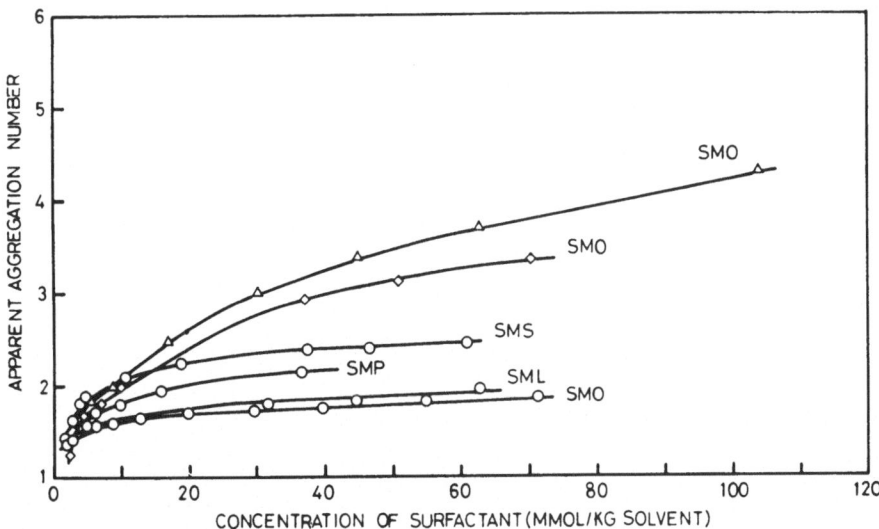

Fig. 18. Relation between apparent aggregation number and concentration of sorbitan monofatty acid esters at 40 °C in different solvents: n-C$_7$H$_{16}$ (△), C$_6$H$_{12}$ (◇), C$_6$H$_6$ (○). [J. Colloid Interface Sci. *49*, 383 (1974)]

oxyethylene 2(1,3-dioctoxypropyl) ether as a nonionic surfactant which was observed not to aggregate in a variety of solvents shown in Table 6 (see also[126]). They found, however, a considerable influence of traces of water with respect to a micelle formation of the nonionics. Later, Kitahara and coworkers[125] investigated again α-monoglycerides and sorbitan monofatty acid esters (sorbitan-monolaurate (SML), -monopalmitate (SMP), -monostereate (SMS), and -monooleate (SMO)) in various nonpolar solvents. The concentration dependence of the apparent aggregation numbers of the latter compounds are shown in Fig. 18, as determined by vapour pressure osmometry. These aggregation numbers appear to be rather small. They should be judged in view of Becher's discussion concerning the comparison between weight and number average molecular weights.

Finally, Cowie and Sirianni[28] investigated some polyoxyethylene – polyoxypropylene surfactants in benzene, dioxane and butylchloride. Again, considerable differences were found between weight and number average molecular weights. This is in agreement with comparable studies of other authors[32, 33, 208].

3.2.1.1 Mixed Solvents

Some investigations on the solvent effect concerning the aggregation tendency of surfactant in nonpolar media have been made in mixed solvent systems. These yield information on the relative stability of a particular micellar structure in one of the solvent components with respect to the other. Also inversion of the micellar structure has been observed in such cases where nonpolar, aprotic solvents were mixed with those giving rise to hydrophilic interactions, i.e., which are in general structure forming.

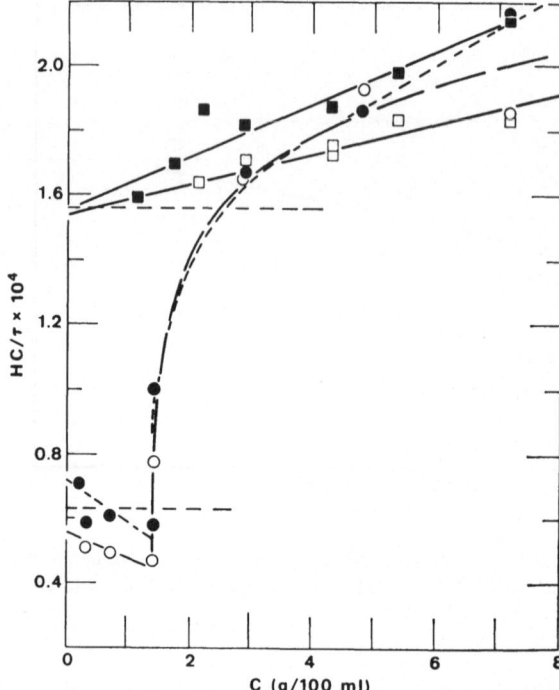

Fig. 19. Lightscattering data for AOT in t-amyl alcohol (square points) and in t-amyl alcohol diluted with nonane (round points). Open and solid points represent Hg-green and blue line data, respectively. [J. Colloid Interface Sci. *29*, 6 (1969)]

Regarding the first case Peri[162] made interesting observations with light scattering experiments by diluting an AOT/t-amyl alcohol master solution with nonane. The latter solvent has almost the same refractive index as t-amyl alcohol (which was essential in view of the lightscattering experiment). Figure 19 shows the Hc/τ-values (which are to a first approximation ~ 1/molecular weight) initially to follow roughly the same plot when nonane was added as on dilution with t-amyl alcohol. Between 60 and 80% nonane, however, Hc/τ decreased rapidly to a minimum. If corrections due to a probable impurity are considered, the final micellar weight corresponds to the value which would have been found at infinite dilution with nonane. This result may lend some support to the possibility that preferred micellar configurations exist. Thus the rather abrupt change in turbidity would indicate a transition between two markedly different micelles.

Similar findings have been reported[147] with copper caprylate in dioxane/ethanol mixtures.

Also Fryar and Kaufman[81] studied the solvent effect on the stability of barium dinonylnaphthalene sulfonate in toluene, toluene/methanol, and methanol solutions by ultracentrifugation and viscometry. The aggregation number of the micelles reduced from about 10 in toluene to about 4 when the mole fraction of free methanol in the solvent mixture was approximately 0.03. In pure methanol BaDNNS micelles did not exist.

An example for the second case mentioned above is reported by Fendler et al.[73] studying hexylammoniumpropionate (HAP) in a polar-nonpolar solvent mixture. Thus the chemical shifts of the magnetically discrete protons of the HAP have been

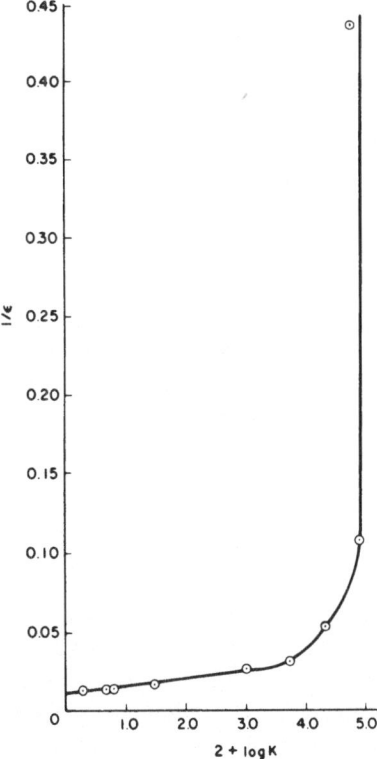

Fig. 20. Dependence of the equilibrium association constant K of HAP (hexylammoniumpropionate) on the reciprocal dielectric constant ϵ of the benzene-d_6-DMSO-d_6 and DMSO-d_6-D_2O solvent system at 23.5 °C. [J. Phys. Chem. *79*, 917 (1975)]

utilized to investigate the association behaviour of this cationic surfactant. Solvent mixtures ranging from 100–0 wt% water to DMSO and 100–0% DMSO to benzene were applied. The aggregation ranged from that of normal to inverted structures (Fig. 20).

Similar investigations were performed by Elworthy and McIntosh[60] using lecithin in suitable water-ethanol-benzene mixtures. In water large micelles of about 6400 monomers are formed, while with decreasing dielectric constant the tendency towards micelle formation decreases. In a 93.4% ethanol – 6.6% H_2O mixture no aggregation occurred. At further decrease of the dielectric constant the conditions become suitable with respect to the formation of reversed micelles (see Fig. 21). In

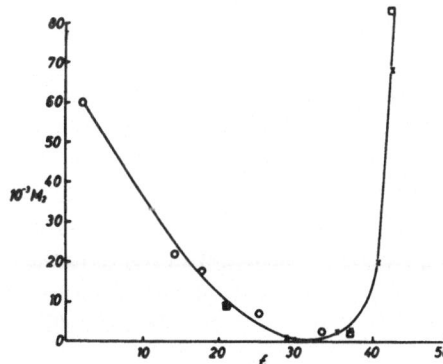

Fig. 21. Micellar weight versus dielectric constant ϵ for natural lecithin. Lightscattering in single (o), (x) in mixed solvents; (□) diffusion-viscosity measurements (mean values). [Kolloid Z. Polym. *195*, 27 (1964)]

spite of this apparent parallelism between aggregational pattern and dielectric constant the latter is hardly an appropriate parameter to describe the solvent effect. Probably Hildebrand's solubility parameter can better serve as a general guide which is related to more relevant molecular properties of the system and hence preferable to predict solvent effects. Investigations with nonionic surfactants (dodecylhexaoxyethyleneglycolmonoether) in water/formamide mixtures have been carried out by McDonald[37] discussing the results with respect to Hildebrand's solubility parameter.

3.2.2 Temperature Dependence

One of the most frequently varied parameters is the temperature of the system: The obvious reason for such measurements was to determine thermodynamic properties of the system, for example, the enthalpy of micellization from the temperature dependence of the critical micelle concentration (CMC) using the well-known relation

$$-RT^2 \cdot \frac{d \ln (CMC)}{dt} = \Delta H_n^{\ominus} = \Delta H_n^{\ominus *}, \tag{15}$$

where $\Delta H_n^{\ominus}$ is the enthalpy change per monomer with respect to the predominantly formed micelle S_n from n monomers. (There exist, particularly in nonpolar surfactant solutions, many examples of reasonably well-defined monodisperse micellar systems). $\Delta H_n^{\ominus *}$ denotes the enthalpy changes per monomer for the addition of a single monomer to a micelle S_{n-1}[150]. It is known that the above relation (15) is not exact unless the aggregation number (n) is independent of temperature. This, however, is true to a first approximation in some cases of micelle in nonpolar surfactant solutions. Only with regard to these examples it might be permissible to apply the above procedure.

There exist, actually, observations that the aggregation number of various surfactant micelles in nonpolar media exhibit a rather weak temperature dependence. Such findings have been reported for ionic surfactants by Peri[162] with AOT in n-nonane between 25 and 75 °C and by Zulauf and Eicke[227] with AOT in isooctane between −20° and +95 °C who could not detect any temperature dependence of the micellar size with the help of very precise photon correlation spectroscopic measurements. These results were confirmed by Ueno and Kishimoto[210] in their investigations of AOT in benzene and cyclohexane between 25° and 45 °C. Also Heilweil[89] found essentially no temperature dependence of the aggregation with sodium 2,6-di-n-dodecyl naphthalene-l- sulfonates in n-decane.

The critical balance between enthalpic and entropic contributions with respect to the stability of micellar aggregates in nonpolar detergent solutions is demonstrated by the temperature effect on the aggregation of sodium tripentylmethylbenzene sulfonate (NaTPMBS) and the corresponding potassium compound[175, 176]. A relative small temperature effect with respect to the micellar size of NaTPMBS in n-heptane compared to KTPMBS was observed. The temperature coefficient with regard to the micelle size, however, depended on the temperature region. The decrease of the association number was considerably larger between 20° and 40 °C than between 40° and 60 °C, indicating, possibly, that a preferred micellar aggregate with increased stability exists[175]. This finding is confirmed by investigations of Strahm[203] and could point to a more pronounced polydispersity. The aggregation of the potassium salt was even more sensitive towards temperature variations (and could be compared in this respect to its above discussed concentration dependence). It was satisfactorily described by a multiple association model with equal association constants. The latter compound exhibited according to the author the association patterns of "typical" cationic surfactants in apolar media. This observation offers the opportunity to point out again that a discrimination of anionic and cationic surfactants according to their aggregation behavior is certainly unjustified. Considering a critical balance between the enthalpic and entropic contributions to the micellization in this particular case, it is conceivable that the reduced hydration interaction of the potassium[46, 228] compared to that of the sodium ion produces this effect. The phosphatidyl cholin (= lecithin) investigated by Elworthy et al.[60, 61] appears also to represent an example of a more pronounced temperature dependence of the micellar size. The apparent micellar weight of lecithin in benzene decreases from 57,000 at 25 °C to about 43,000 at 40 °C.

A careful analysis of the temperature dependence and self-association patterns of dodecylammoniumpropionate (DAP) in benzene and cyclohexane have been conducted by Adams, Fendler et al.[137]. The investigation represents in part a considerable extension of Kreutzer's[130] description of "open" and "closed" aggregation. Several association models were tested. The best fit of the experimental data was obtained with the sequential, indefinite self-association model: the observed temperature dependence corresponded to this result, i.e., decreasing degree of association with an increase of the temperature, where at constant temperature the degree of self-association in cyclohexane was larger than in benzene.

Kitahara[115, 116, 119, 121] arrives at similar conclusions with fatty acid salts of higher aliphatic primary amines in benzene. Large amounts of data on cationic surfactants, particularly, their temperature dependent aggregation were collected by Kertes and coworkers[109, 110, 111, 141]. In a number of cases thermodynamic data were calculated from this temperature dependence[119]. However, frequently the dependence of the aggregation number on the temperature was not duly considered which makes the derived quantities less useful.

Summarizing, as a practical rule cationic surfactants show in general a more pronounced temperature (and concentration) dependent aggregation number.

A quite similar behaviour is found with nonionic surfactants in nonpolar media. Debye and Coll[33] studied the temperature dependent aggregation of a-monoglycerides, in particular, monocaprin (C_{10}) in carbontetrachloride at 23 and 38 °C. They calculated the average heat of association for one mole of monomers which entered into the clusters. Thus, $\Delta H^{\ominus}$ was obtained to be approximately 5.5 Kcal/mole. This value is in the range of hydrogen bond energies. Hydrogen bonding is, indeed, taken to be the essential contribution to the stability of nonionic surfactant aggregates. Simultaneous dipole moment measurements, however, seemed to indicate that no typical reversed micellar aggregate was formed but a more randomly built up cluster.

The same class of surfactants was used by Kitahara[125] to investigate solubility, critical aggregation, micelle formation and its temperature dependence in nonaqueous (apolar) solutions. It was assumed, however, that the so-called phase-separation model could be applied with a fixed aggregation number of the particles. This appears to be less probable with respect to the discussion in the preceeding paragraph. The thermodynamic parameters obtained from this procedure are, therefore, of only minor reliability.

The nonionic surfactants exhibit, similar to "typical" cationic surfactants but to an even more pronounced extent, a considerable temperature (and concentration) dependent aggregation and thus, a remarkable polydispersity of aggregate sizes. Moreover, there seem to exist a large diversity of association patterns which differ considerably in the degree of association. Some types of nonionic surfactants, for example, polyoxyethylene 2 (1,3-dioctoxypropyl) ether, do not all form aggregates in nonpolar solvents according to Kitahara[123]. Corresponding findings were reported by Sirianni et al.[199] and Becher[9].

Micelle formation of the nonionic detergent sorbitan monostearate in o-xylene, particularly the temperature dependence of the CMC, has been observed by Brown et al.[16]. The data were obtained from surface tension, dye solubilization, and light scattering measurements. With regard to the CMC values the results derived from these techniques agreed reasonably well. The ΔH- and ΔS-values evaluated from the remarkable temperature dependence of the CMC cannot claim to be more than an estimate. Two different values of ΔH and ΔS at 25 and 45 °C were determined due to the considerable variation of the heat of micellization with temperature. The discrepancies between the values referring to dye solubilization and surface tension measurements are probably reasonable.

CMC determinations as a function of temperature utilizing the change of the amount of solubilized water are inaccurate. This procedure has been frequently applied. In this way, for example, the effect of the temperature on the CMC of hexaoxyethylene dodecylether in cyclohexane[183] was determined. The CMC loses its well defined meaning in a ternary system, viz. to represent a thermodynamic property of the particular surfactant/solvent system (see Paragraph 2.2).

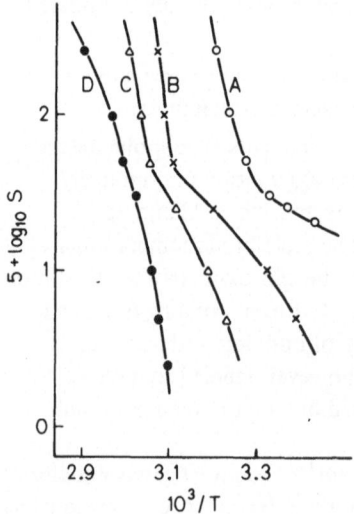

Fig. 22. Solubilities of alkylpyridinium iodides: (•) octadecyl-, (△) hexadecyl-, (x) tetradecyl-, (○) dodecyl- in xylene versus the reciprocal abs. temperature. (J. Chem. Soc. *1956*, 3229)

There exists, actually, another aspect regarding a temperature variation of surfactant solutions: the well-known Krafft-point determination[128]. Since, however, not a micellar property is concerned but the temperature dependence of the monomer activity of the soap molecules, this section is considered more as an appendix to the foregoing discussion.

In comparison to aqueous soap solutions there exist relatively few determinations of Krafft points in nonpolar surfactant solutions. An interesting example is due to Addison and Furmidge[2] who determined the solubilities of four alkylpyridinium iodides (dodecyl-, tetradecyl-, hexadecyl-, and octadecyl-) in xylene (mixture of isomers, b.p. 138 °C). The solubilities were plotted against $1/T$ (see Fig. 22). The breaks in the curves and the chain length effect (which is much less pronounced compared to aqueous solutions) are clearly shown. Similar experiments have been reported by Mehrotra et al.[146] on copper soaps in various organic solvents and by Malik et al.[140] on cobalt hexamine soaps.

4 Kinetics of Micelle Formation in Nonpolar Media

One way to advance beyond purely thermodynamic models and to obtain information on individual steps of the aggregational process is to conduct kinetic experiments. This, however, meets considerable difficulties in nonpolar surfactant solutions. The standard relaxation methods, i.e., temperature[55, 56] and pressure jump[138, 204] techniques are not feasible since their application supposes that the systems are sufficiently conductive (usual procedure with temperature jump technique is to add a 0.1 M inert electrolyte to the aqueous system). Apart from stopped flow techniques which appear to be too slow to follow micelle formation, there exist, essentially, two techniques which are suitable with respect to a kinetic investigation of association processes in "nonconductive" media with low dielectric constants: (i) the ultrasonic absorption[56, 138] and (ii) the dielectric field effect technique[13]. (Modified temperature jump techniques as, for example, suggested by Gerischer et al.[68] might be applicable in some cases but cannot be considered to be a standard method).

It appears justified and necessary to give a slightly more detailed description of the principle of the dielectric field effect technique since Eicke and coworkers[43] applied this method to solutions of AOT in benzene, cyclohexane and dioxane and obtained for the first time kinetic data related to aggregation processes of surfactants in nonpolar solvents.

According to Eigen and coworkers[13] chemical dielectric loss increments can be used to derive kinetic information of reacting chemical systems, if the chemical transformation is accompanied by a finite change of the electrical moment of the system expressed as the molar electric moment of the reaction $\Delta M(E)$ (i.e., the difference of the partial molar moments of the products and those of the reactants). Due to the nonlinear relationship between $\ln K$, the logarithm of the equilibrium constant K of the reacting system, and the electric field strength E, i.e., $\partial \ln K / \partial E = \Delta M / RT$, high static fields modulated by an a.c. field of low amplitude have to be applied to observe finite shifts of the chemical equilibrium. Such a change of the reaction moment is

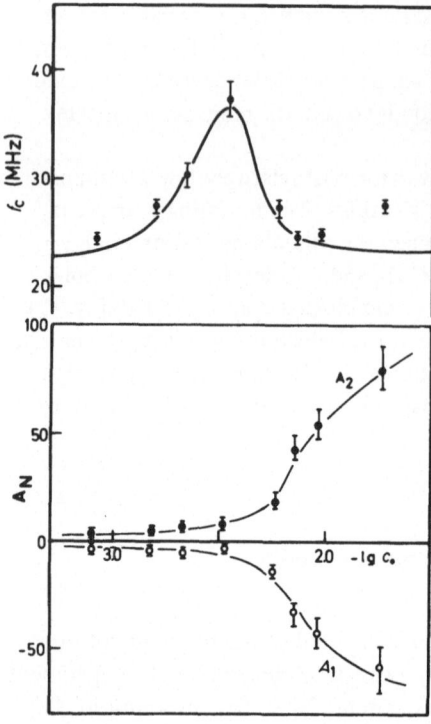

Fig. 23. *Top:* Characteristic frequency f_c versus surfactant (AOT) concentration in cyclohexane, 22.0 °C. Curve through data points calculated according to[43]. *Bottom:* Amplitude factors of the field effect measurements normalized with respect to the applied dc field of AOT/C_6H_{12} solutions: Upper curve (positive amplitude, solid circles): Chemical excess losses. Lower curve (negative amplitudes, open circles): orientational field effect [Ber. Bunsenges. Phys. Chem. *79*, 667 (1975)]

produced by the extensive compensation of the effective dipole moments of the monomers in the (micellar) aggregates[40, 45, 46].

The two diagrams depicted in Fig. 23 appear to characterize the particular features of these systems. The lower diagram shows the amplitude factors of the field effect measurements (normalized with respect to the applied field). The upper curve describes the chemical excess loss, the lower one the orientational field effect. The points of the intersection of two straight lines through the horizontal portions of the curves at low concentrations and the steep slopes at higher concentrations are thought to be identified with the critical micelle concentration of the AOT/cyclohexane system. Comparing both diagrams it becomes apparent that the maximum of the observed chemical relaxation frequencies of the field effect measurements (upper diagram) coincides with the just mentioned points of intersection of the orientational and chemical amplitude curves at $-\log c_0 = 2.5$. It should be realized that an upper limit of the CMC is to be expected due to the field dependent shift of the equilibrium: monomer $\rightleftharpoons$ micelle in favor of the monomers.

The essential result of this investigation is shown in the upper part of Fig. 23. It displays the dependence of the evaluated characteristic chemical frequency $f_c = 1/2\pi\tau$ (τ = relaxation time) on the concentration by keeping the orientational relaxation frequency constant (20 MHz). The curve through the data points was calculated according to the reaction scheme shown below (for more details see[43, 54]). The concentration dependence rules out, apparently, a linear aggregation as the sole relaxation process. In this case, the reciprocal relaxation time versus the concentra-

Table 7. Proposed reaction scheme for topological transformation during micelle formation in nonpolar solvents. AOT in C_6H_{12} [Ber. Bunsenges. Physikal. Chem. *79*, 667 (1975)]

$$3\,C_1 \rightleftharpoons C_3$$

$$C_1 + C_3 \underset{\overline{k_3}}{\overset{k_3}{\rightleftharpoons}} C_3' + C_1$$

$$K_B \Updownarrow \qquad \Updownarrow K_B$$

$$C_1 + C_4 \underset{\overline{k_4}}{\overset{k_4}{\rightleftharpoons}} C_4' + C_1$$

$$K_B \Updownarrow \qquad \Updownarrow K_B$$

$$C_1 + C_i \underset{\overline{k_i}}{\overset{k_i}{\rightleftharpoons}} C_i' + C_1$$

$$K_B \Updownarrow \qquad \Updownarrow K_B$$

$$C_n \underset{\overline{k_n}}{\overset{k_n}{\rightleftharpoons}} C_n'$$

tion would exhibit a curve increasing at the beginning and levelling off at higher concentrations.

It appears interesting to note that quite recently Zana[224] also claimed to have observed (with DAP in cyclohexane applying an ultrasonic relaxation technique) a decrease of the average relaxation frequency with increasing DAP concentration. This particular concentration dependence of the reciprocal relaxation time has been tentatively interpreted with reference to Monod's model[57]. The latter describes conformational changes induced by ligand-enzyme interactions. With regard to the micellization phenomenon it is more appropriate to call this process a topological transformation (L. E. Scriven[185]). Since such a transformation during micelle formation appears reasonable the following reaction scheme has been proposed (see reaction scheme, Tabl 7): C_1 are the monomeric surfactant molecules, C_3 and C_i the trimeric subunits (which have been frequently observed and which were considered as "nuclei": such a concept was confirmed quite recently[46] due to new insights into the role of residual amounts of water with regard to the micellization) and premicellar aggregates (oligomers). In principle, three different elementary steps have to be envisaged regarding the formation of micelles: (i) at very low concentrations $c_0 \ll CMC$, there is an equilibrium between monomers, dimers and trimers. These aggregation processes will be diffusion controlled with bimolecular rate constants in the order of $5 \cdot 10^9$ dm^3 mol^{-1} s^{-1} [13, 94]; (ii) the growing of premicellar aggregates by binding further surfactant monomers at medium concentrations (the aggregation of trimers is less probable, see also[43]); (iii) a topological transformation is to be visualized which will modify the structure of larger premicellar aggregates in such a way that binding of monomers will become too difficult and stop. The surprising monodispersity, recently verified again[227], actually suggests a concept of that kind. Figure 23 demonstrates that the explanation proposed above fits the experimental points quite well.

5 Experimental Methods to Study Aggregational and Micellar Phenomena in Nonpolar Surfactant Systems

The main procedures and essential ideas, both theoretically and experimentally, on which a possible determination of surfactant aggregates in nonpolar solutions are based, originate from comparable investigations in aqueous systems. Accordingly, if a concept regarding such a homo-aggregational process leading to micelles or less defined molecular clusters is proposed, any concentration dependent physical property of the solution sensitive to such a process can be considered, in principle, to be suitable to follow the aggregation of surfactants.

In contrast with aqueous detergent solutions the choice of parameters appropriate to indicate an association process, is rather limited. The reason is found in the (generally) small sized surfactant aggregates formed in nonpolar detergent solutions. Statements as to the size of the micelles can be made if a sufficient "monodispersity" of the particular surfactant system has been proved: i.e., a small micellar weight distribution which deviates not more than about ±10% from the concentration independent apparent molecular weight within the accuracy of the applied method. With this restriction in mind, average (weight and number) apparent micellar weights yield aggregation numbers of not larger than about 40. It cannot be excluded with certainty that the occasionally reported larger aggregation numbers were not sufficiently concentration independent[47]. These aggregates were, therefore, not to be viewed as "micelles" since the concentration independence of their sizes is believed to be an essential requirement regarding their definition.

The up to now most frequently used techniques as, for example, vapour pressure osmometry (VPO) or freezing point depression (with its limitation regarding the solvent dependent measuring temperature) are based upon the colligative properties of the system; the classical absolute light-scattering and ultracentrifugation techniques are only occasionally and approximately applicable with respect to the determination of CMC values. Evaluation of critical micelle concentrations which are based on these latter methods suffer considerably from the insensitivity of these techniques if measurements below the CMC, i.e., below about 10^{-3} mol dm^{-3}, are carried out. More sensitive methods will be discussed below.

In order to avoid repetitions of what is reported in the first part of this review, methods which proved particularly useful in the investigation of nonpolar surfactant solutions will be discussed.

5.1 Methods Based upon Colligative Properties

Both vapour pressure osmometry and depression of the freezing point are the standard techniques probably most frequently used to determine the apparent number average molecular weight of the aggregates. The former method is preferred since the temperature of the sample is easily varied, thus allowing the investigation of the temperature dependence of the aggregate size. The accuracy of the commercially available equipment is rather different, and this has to be carefully considered below

10^{-3} mol dm^{-3} weighed-in concentration. The highest sensitivity which corresponds to the lowest measurable weighed-in concentration of 10^{-4} mol dm^{-3} was achieved with a prototype of a VPO developed by Prof. Simon, ETH Zürich. This corresponds to a measurable temperature difference for benzene of
$\Delta T = ((RT^2)/H_{vap}\,\rho_s) \cdot (c_2/10^3) = 2.1 \cdot 10^{-4}$ °, where H_{vap} is the heat of vaporization (Joule g^{-1}), ρ_s the solvent density (g ml^{-1}), and c_2 the solute concentration in mol dm^{-3}. A careful calibration is necessary with a suitable standard substance structurally similar to the sample to be investigated in order to find out whether any reliable conclusion from measurements in the instrumentally limited concentration region is possible.

There is still another detail which is often not duly considered regarding VPO measurements with nonpolar surfactant solutions: the so-called equilibration time. Tavernier[207] made a careful analysis of the variation of the measured temperature difference with time. He concludes from this theoretical considerations and experimental results (Fig. 24) that extended equilibration times are necessary if surfactant solution are to be investigated. The reason might be seen in the fact that the surface of the solution droplet (within the osmometer) is expected to be partly covered with surfactant molecules. This surfactant "layer" has to be penetrated by the solvent during the equilibration process.

The VPO technique is widely used in connection with methods yielding weight average molecular weight, i.e., light-scattering and ultracentrifugation. From the degree of coincidence between the molecular weights, conclusions as to the monodispersity of the aggregates were deduced. The discrepancies of the aggregation numbers derived from number and weight average determinations puzzled the authors in some cases[32, 33] and led to unwarranted conclusions, particularly with respect to nonionic surfactants in nonpolar solvents. A rather clear account of the solution to this problem was given by Becher[10] who emphasized the different concentration dependences of the apparent weight and number aggregation numbers see Fig. 11. It is seen from this figure that for large aggregation numbers the deviation may become quite considerable if the investigated concentration range is not sufficient

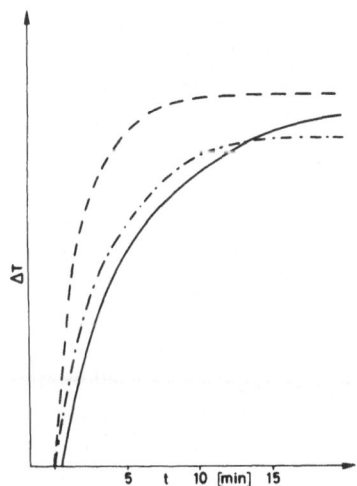

Fig. 24. Temperature variation (ΔT) in VPO measurements versus equilibration time (t). ($---$) methylstereate, (———) iso-hexadecylzincsulfonate, ($-\cdot-\cdot-$) AOT. From: Tavernier, S., Licenciate Univ. Antwerpen (Wilrijk) 1977

within which the VPO measurements were carried out. Such a situation can become critical in studying nonionic surfactants which (in general) form larger aggregates in nonpolar solvents.

The technique has been successfully employed in the investigation of the temperature and solvent dependences of the apparent molecular weights of the aggregates.

5.2 Classical Light-Scattering

A technique which is likewise frequently used in studying aggregation phenomena in nonpolar surfactant solutions is light-scattering. Recommended general treatments regarding classical light-scattering techniques and its applications are found, for example, in[108, 129, 158, 205]. An article, particularly adapted to surfactant systems has been written by Anacker[5].

The experimental procedure is considerably simplified by the modern equipment currently available. Most of the measurements were carried out at a scattering angle of 90° and evaluated by application of the classical Debye equation[22, 31, 96],

$$\frac{Kc}{R_\theta} \left(= \frac{Hc}{\tau} \right) = \frac{1}{M} + 2\,Bc \tag{16}$$

where $K = (2\,\pi^2\,n_0^2/N_A\lambda_0^4)\,(dn/dc)^2$, R_θ the Rayleigh ratio, $\tau = \dfrac{16\,\pi}{3}\,R_\theta$ = turbidity. n_0 is the refractive index of the solvent, λ_0 the wavelength of the incident light, N_A Avogrado's number, dn/dc the differential refractive index between solvent and a solution of concentration c and B is the second virial coefficient.

In applying this technique to the study of micellar solutions the intensity versus concentration function is usually extrapolated to obtain the CMC (at constant scattering angle). Since, however, the CMC occurs in apolar surfactant solutions at concentrations where the turbidity is indistinguishable from that of the pure solvent, extrapolation was frequently made to zero concentration.

Only in those cases where larger particles (aggregates) with higher molecular weights are to be considered, measurements at different scattering angles have to be made. Then the experimental data will be plotted according to the well-known Zimm plot[225], extrapolating the function: $(Kc/R_\theta)_{c,\,\theta}$ versus c and θ, respectively, to zero concentration and zero angle.

Compared to the colligative methods light scattering can yield information on a possible dissymmetry of the aggregates. For anisotropic particles the direction of the electric field associated with the incident light may not coincide with the shift of the electron cloud. The intensity of light scattered at (usually) 90° from anisotropic aggregates is increased over the value predicted on the basis of isotropy by the Cabannes factor.

Although it might appear that this technique is in some respect surpassed by the more powerful and versatile photon correlation spectroscopy (laser beat spectroscopy) which will be discussed in a later paragraph, it can still serve as a convenient method to study the absolute scattered intensity of small colloidal particles.

5.3 Ultracentrifugation

Since its introduction to the study of aggregation phenomena in surfactant solutions by Mathew and Hirschhorn[142] in 1952, the ultracentrifugation technique has become as frequently used as light scattering in order to determine weight average (apparent) molecular weights. This technique was developed to an instrumentation allowing high precision measurements[17, 217] of weight average molecular weights, weight distributions, also diffusion and sedimentation coefficients. The latter are advantageously compared with the corresponding data obtained from photon correlation spectroscopic experiments[227]. Such a comparison between both methods exhibits a disadvantage of the ultracentrifuge method in determining diffusion coefficients (and also apparent molecular weights)[36]: the pressure effect produced in the strong gravitational fields, even at lower speeds, where usually diffusion measurements are carried out. Any light scattering technique is superior if aggregates with only moderate stability are to be investigated. This was shown, for example, in the case of dodecylammonium carboxylates and of pressure sensitive microemulsions in which these surfactants were used to stabilize the system[36].

5.4 Viscosimetry

Determinations of the viscosity of nonpolar surfactant solutions is the last of the classical methods just discussed to evaluate molecular weights. Such a procedure is based upon the relationship

$$[\eta] = \lim_{c \to 0} (\eta_{sp}/c) = K M^{\alpha} \tag{17}$$

where η_{sp} is the specific viscosity $(\eta - \eta_0)/\eta_0$, c the concentration of surfactants in g/100 ml of solution, $[\eta]$ is the intrinsic viscosity being the limiting value of η_{sp}/c at infinite dilution, and M is the apparent molecular weight of the aggregate. These notions can be easily assigned to a micellar surfactant solution, replacing η_0 by η_{cmc}, the viscosity at the critical micelle concentration, and c by c_m, i.e., the weighed-in concentration of surfactant in units of c/n, where n is the aggregation number of micelles. K and α are constants characteristically with respect to the system which are established from viscosity measurements on chemically comparable samples of known molecular weights. For a polydisperse system the average molecular weight M obtained from viscosity measurements is intermediate between the number and weight average molecular weights and is sometimes called the viscosity average molecular weight[169, 205]

$$M_{\eta} = \left(\frac{\Sigma_i n_i M_i^{\alpha+1}}{\Sigma_i n_i M_i} \right)^{1/\alpha} . \tag{18}$$

Using $\alpha = 1$ (Staudinger law) the weight average molecular weight is obtained.

Moreover, the macroscopic viscosity is related according to Einstein[58] and Simha[78] to the volume fraction (ϕ) of the particle in the total volume of the system, i.e., $\eta = \eta_0 (1 + \nu\phi)$ where $\nu = 2.5$ in the case of spheres (Einstein) in the limit of high dilution. This relation is easily shown to yield a connection between the intrinsic viscosity $[\eta]$, the specific volume of the micelles and a form factor ν: $[\eta]$ is thus in principle suitable to render information on the shape of colloidal particles.

Due to the ease of determining viscosities, this technique has been widely applied. It must be emphasized, however, that viscosity data are not in general easily interpreted. If combined with data obtained from other techniques they can be of considerable value in elucidating the underlying physical process. This has been demonstrated, for example, with slightly hydrated micelles in nonpolar media, applying electrical conductivity, dielectric dispersion, and viscosity measurements[20]. In such a case it might be even possible to arrive at a semiquantitative description of viscosity data. Generally, most of the information from viscosimetric measurements refers to phenomenological aspects[99]. This is certainly due to the complexity of multi-body interactions responsible for the viscosity phenomenon. Actually at volume fractions greater than 0.01 the viscosity of a suspension is changed due to the formation of (temporary) doublet, triplet, and higher-order multiplet aggregates which enhance the rate of energy dissipation[88]. Reviews on empirical and theoretical approaches, also relevant to micellar solutions, are given by Rutgers[180] or Frisch and Simha[78].

5.5 Photon-Correlation Spectroscopy (Laser-Beat Spectroscopy)

A method which is basically related to one of the above discussed classical techniques, i.e., light scattering, and which is particularly well adapted to the study of micellar aggregates in nonpolar solutions is the more recently developed photon-correlation spectroscopy[14, 23, 29, 34, 76]. This method yields information on the translational diffusion coefficient of the particles (involved in a fluctuation process) directly from the relaxation of the temporary intensity fluctuation expressed by the intensity-intensity correlation function for the scattered light $G(\tau) = \overline{I(t) I(t + \tau)}$. This expression has to be related to the correlation function which represents the motion of the solute molecules, i.e., $G^*(\tau) = \overline{\delta c(K, t) \delta c(K, t + \tau)}$ where $\delta c(K, t)$ is a measure of the temporal (t) and spatial (wave number K) fluctuations in the solute concentration[76]. The method allows one easily to determine molecular weights by applying the Svedberg equation. Thus molecular weights of very large molecules (aggregates) and those with molecular weights of about 100 are accessible with this technique. It appears noteworthy that it is in principle possible to obtain information on rotational diffusion coefficients from the self beating spectrum (the most frequently applied technique). In this case it is necessary to observe a (usually small) polarized component of the scattered light perpendicular to that of the normal scattering. Mazer, Benedek et al.[143, 144, 145] in the US, Corti and Degiorgio[29] in Italy have pioneered this method regarding its application to micellar phenomena in aqueous solutions.

Although there exist already a number of adequate investigations in aqueous detergent solutions[27, 95, 143, 144] hardly any study has been carried out in nonpolar surfactant systems. Only recently, Zulauf and Eicke[227] successfully applied the photon correlation spectroscopy to investigating AOT micelles in isooctane and the ternary W/O-microemulsion (Water/AOT/isooctane). Remarkable results were obtained.

Inverted AOT micelles were found to be stable between −20 and +95 °C (even at −85 °C no turbidity was detected). The Stokes radius of the AOT micelles was determined to be 15 ± 0.3 Å (1.5 ± 0.03 nm) independent of the weighed-in concentration (8 10^{-3} to 2 10^{-1} mol dm^{-3}), and independent of the scattering angle. It was not possible, however, to determine the CMC due to its small value in isooctane (~ 5 10^{-4} mol dm^{-3}).

Experimental evidence was presented, moreover, that a clear distinction is possible between micellar and (W/O) microemulsion regions according to the amount of solubilized water. Also, an optical matching phenomenon was observed which is of considerable significance regarding an understanding of the molecular properties of microemulsions in general.

Thus it appears justified to expect that this technique will prove invaluable for investigations of nonpolar detergent solutions and even more for the study of phase transitions (at higher surfactant concentrations).

5.6 Fluorescence/Nanosecond Fluorescence Spectroscopy

The methods discussed so far are in many instances due to their low sensitivity at small weighed-in concentrations in nonpolar surfactant solutions not too well suited to determine critical micelle concentrations. This well-known difficulty led Kaufman and Singleterry[103, 104] to suggest the solubilization of a fluorescent dye molecule, for example, Rhodamine B in order to detect reversed micelles at very low detergent concentrations. The transition from monomers to micellar aggregates can then be followed by the depolarization of the fluorescence. The essential drawback of this method which reduces its value with regard to a CMC determination is the effect of a third component on the critical micelle concentration[8, 122]. The solubilization of a dye by the soap molecules is the obvious manifestation of a modification of the system. Such a solubilization is equivalent to a stabilization of the system and causes (at least in principle) a shift of the CMC to smaller weighed-in concentrations[48]. Hence with CMC values in the order of 10^{-7} to 10^{-6} mol dm^{-3} [103] this restriction should be kept in mind.

In contrast to the evaluation of critical micelle concentrations the determination of the size of micellar aggregates utilizing the fluorescence depolarization is reasonable as long as the solubilization does not affect the (average) degree of association of the particles and their shapes. Use is made of Perrin's relation[164, 196] between the volume V of the micelle and the polarization of light emitted by an immobilized dye molecule p_0 and that emitted by the dye solubilized in a micelle p, i.e.,

$$V = \left(\frac{1}{p_0} - \frac{1}{3} \right) \frac{\tau\, RT}{\eta} \frac{p\, p_0}{p_0 - p} \tag{19}$$

where τ is the average excited lifetime of the dye molecule and η the viscosity of the solution.

Micelle volumes estimated from fluorescence depolarization measurements have been found to agree well with those computed from osmotic pressure determinations

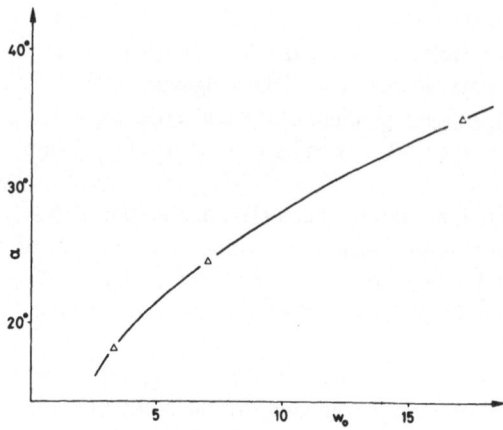

Fig. 25. Angle between emission and absorption dipole in nanosecond fluorescence measurements versus solubilized water ($w_0 = [H_2O]/[AOT]$). [J. Colloid Interface Sci. *65*, 131 (1978)]

of soap solutions[196]. The apparent advantage of this method is that 10^{-8} mole per dm^3 of dye furnishes sufficient fluorescence intensity for useful measurements. A typical plot obtained with this method by Kaufman and Singleterry[103, 105] of the apparent aggregation number versus stoichiometric concentration for sodium and barium dinonylnaphthalene sulfonates, respectively, in benzene are shown in Fig. 2.

Probing of reversed micelles with dye molecules and observing the fluorescene polarization has recently again received considerable interest. It has been realized that even more information is obtainable from fluorescence measurements compared with the above discussed investigations with reversed micellar systems. Thus, Wong et al.[221, 223] derived data concerning the so-called "microviscosity" of the interior of AOT reversed micelles studied by fluorescence polarization. The emission of rhodamine B is strongly polarized in the "absence" of water (i.e., below the detectable amount[46]) indicating a very rigid structure of the micellar core. Also the emission of pyrene sulfonic acid fluorescence in the presence of quenchers (Cu^{2+}, Tl^+) showed that the motion of the ionic quenchers is rather restricted at small water core sizes.

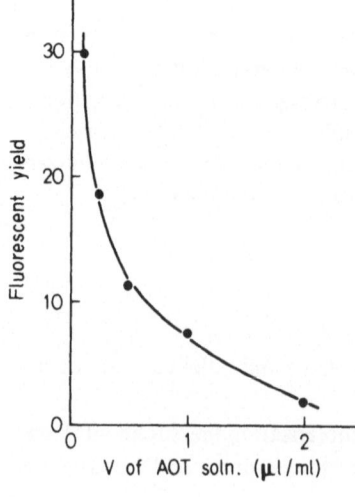

Fig. 26. Increased fluorescent yield (arbitrary scale) per Tb^{3+} ion at 545 nm as produced by decreasing the volume of solubilizate in each micelle. V = volume of mixture (1:1 by volume) of 0.5M $TbCl_3$ and 0.5M HPA (= hydroxyphenylacetic acid) solutions which has been solubilized in 1 ml of $2 \cdot 10^{-2}$mol dm^{-3} AOT in isooctane. [J. Colloid Interface Sci. *56*, 168 (1976)]

The nanosecond fluorescence spectroscopy (or photon sampling technique[15, 216])
is a very suitable tool for investigating reversed micelles. Again, an appropriate dye
molecule has to be solubilized in the polar core of the aggregates. The advantage of
this technique is beyond the conventional measurements of the depolarization of
the dye fluorescence due to the possibility of determining the angle between emission
and absorption dipoles (Fig. 25). It is thus feasible to obtain information on the mobility of the surfactant molecules constituting the micellar "membrane". Such experiments have been done by Eicke and Zinsli[50, 226] using sodium fluorescein solubilized in reversed AOT micelles in isooctane with varying amounts of AOT and water.

Fluorescent probes in reversed micellar aggregates proved, moreover, appropriate
to study dynamic properties of micelles in nonpolar solvents[42, 44]. The particularly
suitable fluorescent label was the highly water sensitive terbium ion (Tb^{3+}) (Fig. 26).
A rapid exchange of electrolyte solutions and water between AOT micelles in isooctane occurring during collisions of micelles was evidenced. A theoretical analysis
of the exchange mechanism supports the idea that fast kinetic processes are involved.

5.7 NMR-Techniques

The application of the nuclear magnetic resonance spectroscopy, generally NMR, to
surfactant solutions, in particular, nonpolar detergent systems has proved very successful. In some more recent investigations also other nuclei as, for example, sodium, lithium, carbon and fluorine were advantageously used. The techniques are quite extensively treated in the first part of this review. Therefore, only a few supplementary details with respect to nonpolar surfactant systems have to be added. As in aqueous systems, chemical shift, line width, and spin-lattice relaxation time (τ_1) have been
generally related to the hydration interaction between water and the polar (ionic)
groups of the detergent molecules in order to study their structures and stabilities.

Such investigations have been carried out, for example, by Oedberg et al.[201] on
anionic surfactants (lithium and cesium dinonylnaphthalene sulfonates) in heptane,
by Wong et al.[222] on AOT in heptane using ^1H and ^{23}Na NMR, by Fendler and Fendler[70, 71] on alkylammonium carboxylate micellar aggregates in various nonpolar solvents, by Walter et al.[215] and Fung et al.[85] on phosphatidylchloline micelles in carbon tetrachloride, and by Gentile et al.[86] on structure and hydration of nonionic
inverted detergent micelles.

In recent years ^{13}C spin-lattice relaxation time measurements have been often
applied which can provide quantitative information on the surfactant mobility in aggregates (see first part). Ueno et al.[209, 211] studied with the help of this technique
and lanthanide shift reagents the association of AOT in chloroform. These authors
concluded that the mobility of the side chains of the hydrocarbon moiety (2-ethyl-
hexyl) is more restricted than that of the main (= hexyl) chain. A more detailed analysis has been performed by Denss[36] (see also[50]) with AOT in benzene-d_6. With
regard to the correlation time τ_c it was assumed to a first approximation that the
correlation time of the ^{13}C-nuclei is determined by magnetic dipole-dipole interaction
between ^{13}C and ^1H nuclei. If the molecular rotation is anisotropic and/or rotation
along the C-C bonds are occurring then τ_c has to be considered as an average correlation time. Denss' molecular interpretation followed the work of Williams et al.[218],

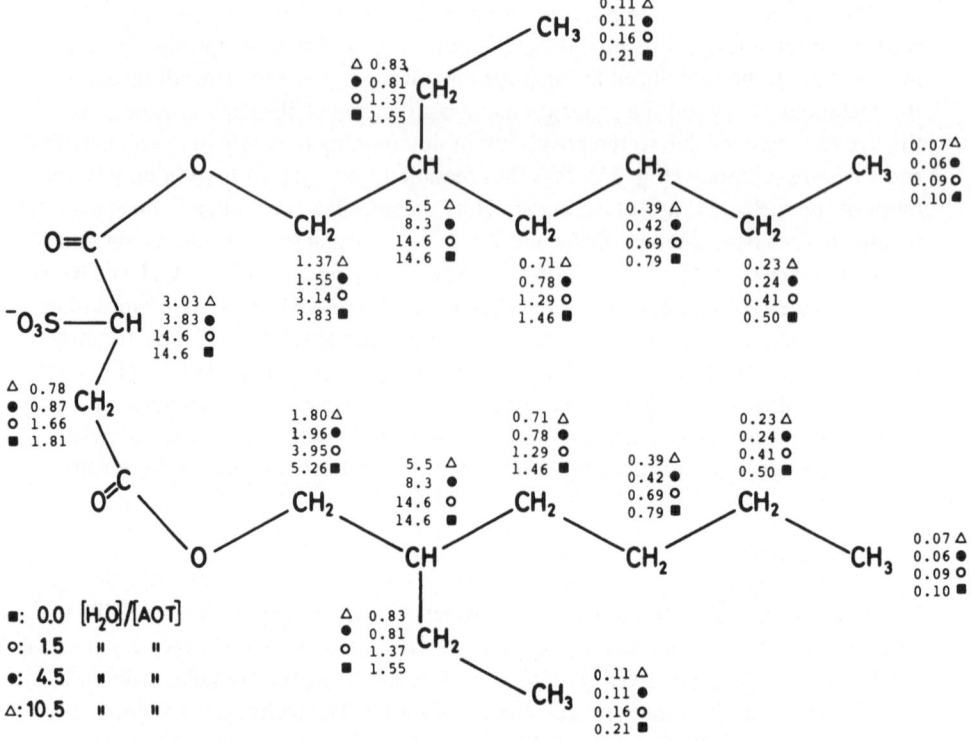

Fig. 27. Variation of correlation time $\tau \cdot 10$ by different amounts of solubilized water $(0 \leq [H_2O]/[AOT] \leq 10.5)$ in 0.8 M AOT solutions in C_6D_6. ^{13}C-NMR measurements, $H_0 = 23.5$ kG, 25 °C.
from: Denss, A., Ph D thesis, Univ. Basel 1977 [see also (50)]

Allerhand et al.[3], and Doddrell et al.[35]. These authors concluded that a steady change of τ_c along a flexible molecular chain indicates a motional restriction of one end of the chain. The mobility increases with decreasing τ_c-values, see Fig. 27.

A disadvantage of the PMR technique, in particular, if applied to nonpolar surfactant systems is its low sensitivity. Although this has been considerably improved since the introduction of the Fourier transform technique it is still difficult to study very low concentration regions and, thus, to detect reliably critical concentrations.

5.8 Dielectric Increment and Dispersion Measurements

It appears that relative few dielectric studies of nonpolar detergent solutions have been performed, and these more recently, although this technique is particularly suited to investigate aggregational processes. Whether or not information on aggregates smaller than micelles can be gained depends upon the sensitivity of the differential dielectric bridges or dipole meters. The sensitivity of the commercially available dipole meters are in the order of $\Delta\epsilon/\epsilon = 10^{-5}$ at a frequency of about one Mega-Hertz. Many surfactants, especially ionic detergents, frequently possess rather

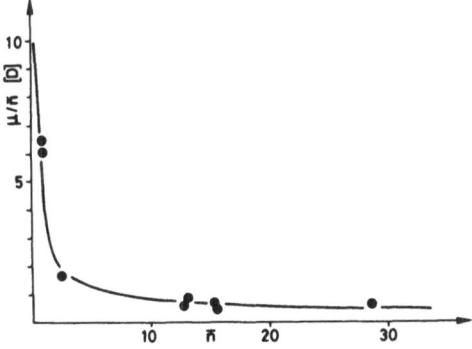

Fig. 28. Apparent dipole moments per monomer versus aggregation number of AOT in different nonpolar solvents. [J. Colloid Interface Sci. *48*, 281 (1974)]

large dipole moments. In the course of the aggregation process the overall dipole moment of the molecular cluster changes due to a mutual compensation of the dipole moments of the individual soap molecules[40], see Fig. 28. According to this figure the initial change of the apparent dipole moment per monomer $\mu/\overline{n}$ is easily detectable. The dielectric increment method should, therefore, be suitable to determine critical micelle concentrations. This technique has been applied recently by Eicke and Christen[46] again to nonpolar solutions of AOT and DAP in benzene and cyclohexane. Figure 29 shows after an initial increase in the dielectric increment a sudden drop of $\Delta\epsilon$ which is then followed by an increase of the dielectric increment with the weighed-in concentration of detergent. The concentration corresponding to the break of the dielectric increment versus concentration curve coincides very satisfactorily with comparable findings obtained by other techniques. The initial increase of the increment, moreover, may give hints as to possible premicellar structures. Since the dielectric measurements yield only indirect information on the formation of micellar aggregates, it is necessary to consult other methods in order to identify the above mentioned break of $\Delta\epsilon$ (c) with a CMC.

If dielectric dispersion measurements have been performed, i.e., the dependence of the dielectric increment on the frequency has been determined, a decision can be made in favor of or against the formation of closed aggregates. Such investigations have been conducted with Aerosol AY (= sodium di-pentylsulfosuccinate) in benzene and varying amounts of solubilized water and aqueous electrolytes by Eicke and Shepherd[41]. The complex permittivity has been observed in the frequency range of 200 kHz to 10 MHz. Mean relaxation times $\overline{\tau} = 4\,\pi\eta \cdot r^3/kT$, (where η = the viscosity of the solution and r = radius of the spherically assumed particle) have been determined from a Cole-Davidson plot where the so-called spread parameter ($\beta = 1$ for a single Debye dispersion) yields additional information concerning the deviations from a spherical shape of the aggregates. If data with respect to the shape of the colloidal particles are available, Perrin's[165] equations describing the rotational diffusion coefficient of prolate or oblate ellipsoidal particles may be used to obtain a more detailed picture of the geometrical shape of the micellar aggregates.

A particular refined dielectric method is based upon the so-called dielectric field effect (or chemical dipole field effect) suggested by Bergmann, Eigen, and DeMaeyer[13]. High static fields modulated by a low frequency a.c. field are applied.

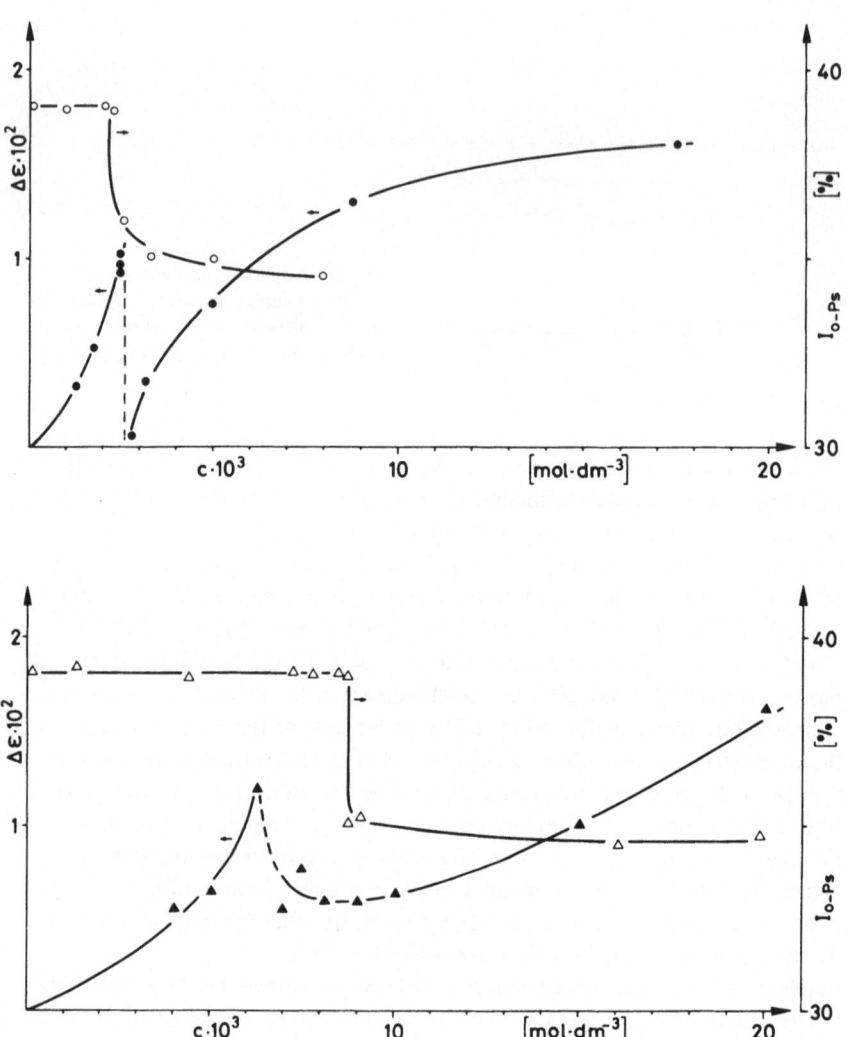

Fig. 29. Dielectric increment ($\Delta\epsilon$) and intensity of o-Positronium ($I_{o\text{-}Ps}$) versus surfactant concentration (top: AOT/C_6H_6, bottom: DAP/C_6H_6) at 25 °C. [Helv. Chim. Acta *61*, 2258 (1978)]

The former, in order to observe finite shifts of the equilibrium between two different polar states and the second superimposed a.c. field serves to measure the expected dielectric loss increments (produced by the field induced shift of the chemical equilibrium). Also, loss decrements are observed due to the considerable alignment of the particles in the high static field which alters the Boltzmann distribution. If both of these effects (which possess opposite signs) occur in the same frequency region, they will be superimposed on each other. Hence, the total dielectric loss is given by

$$\Delta \tan \delta = (\Delta \tan \delta)_{\text{max}}^{\text{ch}} \cdot \phi'' \left(\tau_{\text{ch}} \right) - (\Delta \tan \delta)_{\text{max}}^{\text{or}} \cdot \phi'' \left(\tau_{\text{or}} \right) \tag{20}$$

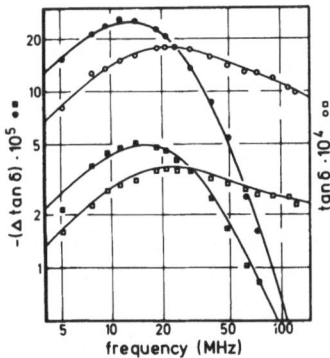

Fig. 30. Orientational relaxation (without applied field: open data points, *right ordinate*) and orientational field effect ($E = 2 \cdot 10^7 \, Vm^{-1}$, filled data points, *left ordinate*) of AOT in $C_6H_{12} \cdot c_0 = 0.0182 \, mol \, dm^{-3}$ (*circles*), $0.0034 \, mol \, dm^{-3}$ (*squares*) 22.0 °C. [Ber. Bunsenges. Phys. Chem. *79*, 667 (1975)]

where $\phi''(\tau) = f/f_c(1 + f^2/f_c^2)$, f_c is the characteristic frequency associated with the amplitude factor $(\Delta \tan \delta)_{max}^{ch} = \Gamma \Delta M^2/(2 \, \epsilon_0 \epsilon_r RT)$, $\Gamma = (\sum_i v_i^2/n_i)^{-1}$, n_i = amount of substance i, v_i = stoichiometric coefficient of i-th reactant, ϵ_0, ϵ_r are the permittivity of the vacuum and the relative permittivity of the solution, ΔM is the molar reaction moment = $\mu_{mic}^2 - \mu_{monomer}^2$, and f is the running frequency[43]. As an example, measurements of the orientational relaxation (without applied electric field) and orientational field effect of AOT in cyclohexane are shown in Fig. 30.

5.9 Positron Annihilation Technique

This technique represents a novelty among the many methods used to investigate nonpolar detergent solutions. It appears that, similarly to the dielectric increment determinations, the system is not disturbed by the measurements and is, therefore, particularly suited to detect characteristic parameters of the system. This technique[1], already well-known in other areas of physical chemistry, has recently been introduced by Ache and coworkers[100, 101] to the study of aqueous and nonpolar micellar systems.

Since it is a relatively new technique in the field of detergent solutions, it appears worthwhile to give a short introduction into the principle of this method (see, for example[138]): Positrons produced by a radioactive decay of neutron deficient nuclides (for example, ^{22}Na) and thermalized by collisions with surrounding matter (within about $7 \cdot 10^{-10}$ s), can abstract electrons from the latter and form the bound state of a positronium (Ps). Two ground states exist: the singlet (para-Ps) has a self-annihilation lifetime in free space of $1.25 \cdot 10^{-10}$ s and decays by a two photon emission. The triplet (ortho-Ps) has a lifetime of $1.4 \cdot 10^{-7}$ s. Self-annihilation occurs via a three photon emission. Positronium lifetime distributions are obtained by standard techniques[219]. The lifetime spectra are separated into a short-lived component, i.e., free positron annihilation and a long-lived component with a decay constant λ_2 and an associated intensity I_2.

Since lifetime and annihilation characteristics of a positron and the formation of the positronium atom (Ps) are determined by their microscopic chemical and physical environment[1, 148)] the positron annihilation technique is expected to be sensitive towards structural changes: it could be shown, actually, that the positronium formation reflects most sensitively structural variations in liquid crystals[157)]. Variations of the mesomorphic phases clearly changed the positron lifetime. Thus, an application to micellization processes suggested itself, since also cooperative, size-limited, aggregational processes show features similar to classical phase transitions.

Therefore, Ache and coworkers studied positron interactions in aqueous and nonpolar detergent systems, i.e., DAP in benzene, cyclohexane, n-hexane, and AOT in benzene. They obtained a rather abrupt drop of the intensity I_2 which is a remarkably sensitive indication of critical concentration, as compared with the information obtained from spectroscopic or other standard techniques. The method appears particularly promising regarding CMC determinations in nonpolar surfactant solutions. It should be realized, however, that the parameters which control the positronium formation process and, consequently, I_2 are not unambiguously known (see[101)] and the literature cited therein). Apparently, the positronium formation is very sensitive to any structural change in the physicochemical environment. According to Cole and Walker[24)] the lifetime of o-Ps seems to be particularly sensitive to changes of intermolecular dipole-dipole interactions as observed, for example, with phase transitions in liquid crystals. A similar mechanism could be responsible of the changes of I_2 due to the aggregation of DAP and AOT in nonpolar solvents.

Very recent studies[52)] compared dielectric increment, $\Delta \epsilon$ with I_2 changes in the system DAP and AOT in benzene, see Fig. 29. The coincidence of the breaks in the corresponding curves is remarkable. It appears, however, that the drop in the $\Delta \epsilon$-curve is more distinct in the AOT/benzene system as compared with DAP/benzene. It should be noted that the micelle formation of AOT in nonpolar solvents is primarily determined by hydration interactions between sodium ions and water molecules and the concomitant hydrogen bond formation linking the surfactant molecules. This process leads to a stable hydrogen bond network and consequently to the micelles. This micelle formation is accompanied by a partial compensation of the individual dipole moments of the soap molecules. Also in the case of DAP a dipole compensation interaction of the ammonium compared to the sodium ion and the correspondingly inferior bond formation the tendency towards structure formation of these aggregates is expected to be less pronounced. Moreover, structural variation, should be considered comparing these two surfactants.

In spite of many favorable examples of coincidence between the conventionally defined CMC-values obtained by this method and, for example, spectroscopic or other techniques, it appears a priori not advisable to assign these breaks directly to critical micelle concentrations without further analyses of these systems.

6 Remarks on the Solubilization and Microemulsion Phenomena

A review on micelles is not complete without mentioning at least a phenomenon which emerges from the very same surfactant property as micellization does, i.e., the so-called " solubilization". It describes the dissolving of a solid, liquid or gas by

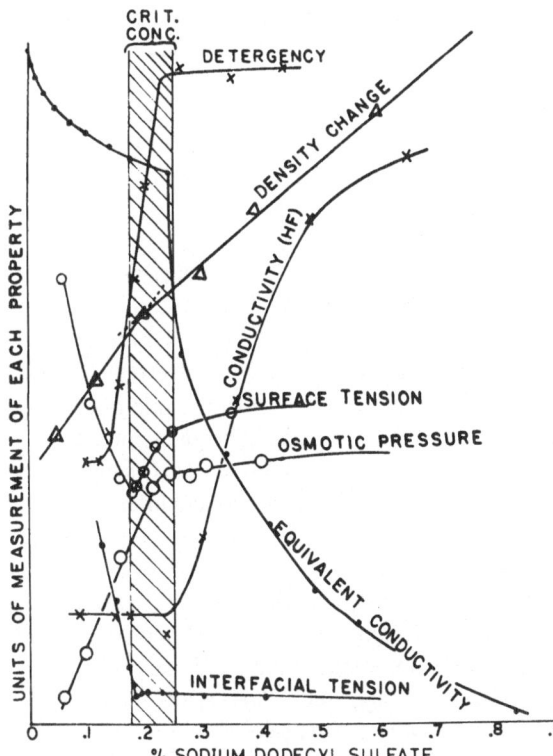

Fig. 31. Changes in some physical properties of an aqueous solution of sodium dodecylsulfate in the neighborhood of the CMC. [J. Phys. Colloid Chem. *52*, 84 (1948)]

an interaction with surfactant molecules. These detergent molecules cover the solubilizate so as to shield it from the respective solvent, thus forming a thermodynamically stable isotropic solution. Solubilization and micellization are, accordingly, competitive processes. This is nicely demonstrated by Fig. 31 where the onset of the "detergent" (= solubilization) process clearly precedes that of the micellization. The start of the latter (as indicated by the breaks in the curves of the corresponding physical properties of the system) is denoted by the shaded area. This experimental finding has been confirmed theoretically[48, 214] by showing that the surfactant concentration necessary to form a thermodynamically stable emulsion, i.e., a microemulsion, is always smaller than the critical micelle concentration in the respective dispersion medium (= solvent). It is probably hardly necessary to mention again that this point deserves particular attention with regard to the preceding discussion on CMC determinations in nonpolar surfactant solutions.

In spite of the intimate connection between micellization and solubilization, i.e., the formation of so-called microemulsions and its industrial importance[4], it is quite impossible referring and discussing the interesting but voluminous material which has been collected up to now. This information is covered in part by some more recent reviews, for example, by Shinoda et al.[193, 194], Fendler and Fendler[72], Shinoda and Friberg[195], Prince[171], Friberg[79, 80], Schick[184], Sherman[189], Lissant[133], and in relation to pharmaceutical and biological applications by Elworthy, Florence and MacFarlane[62].

Only some more general points are to be made in this paragraph. In principle the formation of the two antagonistic types of microemulsions, i.e., oil in water and water in oil, appears to be similar: this follows not only from symmetry considerations but also from the observed "phase inversion", i.e., a transition from a W/O- to a O/W-type microemulsion[195]. The controversy regarding the nature of microemulsions, i.e., whether they are closely related to micellar solutions[195] or basically different from micelles[171, 172, 186] may possibly be settled from new experimental information[227]. Photon-correlation spectroscopic investigations exhibited a reproducibly, well-defined transition between a micellar solution containing solubilized water up to about 10 moles per mole surfactant (system: $H_2O/AOT/i-C_8H_{18}$) and the W/O microemulsion region. The latter region differs clearly from the former by its highly sensitive temperature dependence. The micellar region of inverted micelles is stabilized by hydration interactions[46]. The latter merges into the microemulsion region with increasing amounts of solubilized water. The phenomenologically observed different pattern of micelles and microemulsions are thus essentially due to the considerable energy gap between hydration interaction and interfacial free energies.

There exist certainly a large number of parameters which influence or even determine the extent of the solubilization as, for example, the structure of the surfactant, the physico-chemical property of the solubilizate (in the case of W/O-microemulsions, e.g., aqueous electrolytes), temperature, pressure, electric fields, and, particularly, the so-called cosurfactants[182]. Recently it has been suggested[49] (see also[187]) that one can describe the microemulsion phenomenon from a more general thermodynamical and statistical view point. Such considerations can be traced back to thermodynamic treatments of ternary systems[170], which appear suitably to generalize and unify many of the experimental observations made in connection with microemulsions. An especially interesting phenomenon (which is also of eminent practical importance) is the above mentioned cosurfactant effect on the solubilizing power and stability of microemulsions. Even in this case a thermodynamic approach could be proposed, the issue of which is a generalization of many detailed observations in order to gain a clearer insight into the relations connecting various parameters which determine the experimentally accessible quantities[51].

7 References

1. Ache, H. J.: Angew. Chem. Intern. Edit. *11*, 179 (1972)
2. Addison, C. C., Furmidge, C. G. L.: J. Chem. Soc. *1956*, 3229
3. Allerhand, A.: J. Am. Chem. Soc. *93*, 544 (1979)
4. Al-Rikabi, H., Osoba, J. S.: Paper presented at ACS Meeting. June 9, 1973. Dallas, Texas
5. Anacker, E. W.: In: Cationic surfactants. p. 203. Jungermann, E. (ed.). New York: Marcel Dekker 1970
6. Arkin, L., Singleterry, C. R.: J. Colloid Sci. *4*, 537 (1949)
7. Arnold, V.: PhD Thesis, Univ. Basel 1974
8. McBain, J. W., Merrill, R. C., Vinograd, J. R.: J. Am. Chem. Soc. *62;* 2880 (1940)
9. Becher, P.: J. Phys. Chem. *64*, 1221 (1960)
10. Becher, P.: Nature *206*, 611 (1965)

11. Becher, P.: Micelle formation in aqueous and nonaqueous solutions. In: Nonionic surfactants. Schick, M. J. (ed.). New York: Dekker 1967
12. Benedeck, G. B.: Thermal fluctuations and scattering of light. In: Statistical Physics. Vol. 2. Chretin, M. Gross, E. P., Deser, S. (eds.). New York: Gordon & Breach 1968
13. Bergman, K., Eigen, M., De Mayer, L.: Ber. Bunsenges. Phys. Chem. *67*, 819 (1963)
14. Berne, B. J., Pecora, R.: Dynamic light scattering. New York: Wiley 1976
15. Binkert, T., Tschanz, H. P., Zinsli, P. E.: J. Lumin. *5*, 187 (1972)
16. Brown, C. J., Cooper, D., Moore, J. C. S.: J. Colloid Interface Sci. *32*, 584 (1970)
17. Chervenka, C. H.: Manual of methods for the analytical ultracentrifuge. Publ. by Spinco, Div. of Meckman Instr. 1969
18. Christen, H., Eicke, H. F., Jungen, M.: Helv. Chim. Acta *56*, 216 (1973)
19. Christen, H., Eicke, H. F.: J. Phys. Chem. *78*, 1423 (1974)
20. Christen, H., Eicke, H. F., Rehak, J.: Equilibrium considerations of micelle formation in apolar media in the presence of solubilized water. Proc. Intern. Conf. Colloid a. Surface Sci. Vol. 1, p. 481 Wolfram, E. (ed.). Budapest: Akademiai Kiado 1975
21. Christen, H., Eicke, H. F.: Ber. Bunsenges. Phys. Chemie 1979 (in preparation)
22. Chu, B.: Molecular forces, New York: Interscience 1967
23. Chu, B.: In: Laser light scattering. New York: Academic 1977
24. Cole, G. D., Walker, W. W.: J. Chem. Phys. *42*, 1692 (1965)
25. Conway, B. E.: Electrochemical Data. Amsterdam: Elsevier 1952
26. Copenhafer, D. T., Kraus, C. A.: J. Am. Chem. Soc. *73*, 4557 (1951)
27. Corti, M., Degiorgio, V.: Chem. Phys. Lett. *53*, 237 (1978)
28. Cowie, J. M. G., Sirianni, A. F.: J. Am. Oil Chem. Soc. *43*, 572 (1966)
29. Cummins, H. Z., Pike, E. R. (eds.): Photon correlation and light beating spectroscopy. New York: Plenum 1974
30. Czerlinski, G., Diebler, H., Eigen, M.: Z. Phys. Chem. *NF 19*, 246 (1959)
31. Debye, P.: Ann. N. Y. Acad. Sci. *51*, 575 (1949)
32. Debye, P., Prins, W.: J. Colloid Sci. *13*, 86 (1958)
33. Debye, P., Coll, H.: J. Colloid Sci. *17*, 220 (1962)
34. Degiorgio, V.: Alta Frequenza, XLI, 787 (1972)
35. Doddrell, D., Allerhand, A.: J. Am. Chem. Soc. *93*, 1558 (1971)
36. Denss, A.: Ph D Thesis, Univ. Basel 1977
37. McDonald, J.: J. Pharm. Pharmacol. *22*, 148 (1970)
38. Eicke, H. F., Arnold, V.: J. Colloid Interface Sci. *46*, 101 (1974)
39. Eicke, H. F., Christen, H.: J. Colloid Interface Sci. *46*, 417 (1974)
40. Eicke, H. F., Christen, H.: J. Colloid Interface Sci. *48*, 281 (1974)
41. Eicke, H. F., Shepherd, J. C. W.: Helv. Chim. Acta *57*, 1951 (1974)
42. Eicke, H. F.: Chimia *31*, 265 (1977)
43. Eicke, H. F., Hopmann, R. F. W., Christen, H.: Ber. Bunsenges. Phys. Chem. *79*, 667 (1975)
44. Eicke, H. F., Shepherd, J. C. W., Steinemann, A.: J. Colloid Interface Sci. *56*, 168 (1976)
45. Eicke, H. F.: Micelles in apolar media. In: Micellization, solubilization and microemulsions. Vol. 1, p. 429, Mittal, K. L. (ed.). New York: Plenum 1977
46. Eicke, H. F., Christen, H.: Helv. Chim. Acta *61*, 2258 (1978)
47. Eicke, H. F., Denss, A.: J. Colloid Interface Sci. *64*, 386 (1978)
48. Eicke, H. F.: J. Colloid Interface Sci. *52*, 65 (1975)
49. Eicke, H. F.: J. Colloid Interface Sci. *59*, 308 (1977)
50. Eicke, H. F., Zinsli, P. W.: J. Colloid Interface Sci. *65*, 131 (1978)
51. Eicke, H. F.: J. Colloid Interface Sci. (1979) in press
52. Eicke, H. F., Denss, A.: Croat. Chim. Acta (1979) in press
53. Eicke, H. F., Rehak, J.: Helv. Chim. Acta *59*, 2883 (1976)
54. Eicke, H. F., Christen, H., Hopmann, R. F. W.: Stability considerations with respect to conformational transformations during micelle formation in apolar media. In: Proceedings internat. conference on colloid and surface science. Vol. 1, p. 489 Wolfram, E. (ed.). Budapest: Akadémiai Kiadó 1975
55. Eigen, M.: Disc. Faraday Soc. *17*, 194 (1954)

56. Eigen, M., De Maeyer, L.: Theoretical basis of relaxation spectrometry. In: Investigation of rates and mechanism of reactions. Part II. Hammes, G. G. (ed.). Techniques of chemistry. Vol. VI. New York: Wiley-Interscience 1974
57. Eigen, M.: Quarterly Rev. Biophysics *1*, 3 (1968)
58. Einstein, A.: Investigations on the theory of Brownian movement. New York: Dover 1956
59. El Seoud, O. A., et al.: J. Phys. Chem. *77*, 1876 (1973)
60. Elworthy, P., McIntosh, D. S.: Kolloid-Z. Z. Polymere *195*, 27 (1964)
61. Elworthy, P. H.: J. Chem. Soc. *1959*, 813
62. Elworthy, P. H., Florena, A. T., Maefarlane, C. B.: Solubilization by surface active agents. London: Chapman & Hall 1968
63. Ekwall, P.: J. Colloid Interface Sci. *29*, 16 (1969)
64. Ekwall, P., Mandell, L. Fontell, K.: J. Colloid Interface Sci. *33*, 265 (1970)
65. Ekwall, P., Danielsson, I., Stenius, P.: Aggregation in surfactant systems. In: Surface chemistry and colloids. MTP international review of science. Vol. 7, p. 97, Kerker, M. (ed.) London: Butterworths (1972)
66. Ekwall, P.: In: Adv. in liquid crystals. Vol. I, Brown, G. (ed.). New York: Academic 1975
67. Emeleus, H. J., Anderson, J. S.: Modern aspeczs of inorganic chemistry. London: Routledge & Kegan 1960
68. Ertl, G., Gerischer, H.: Z. Elektrochem. *65*, 629 (1961)
69. Fendler, J. H.: Accts. Chem. Res. *9*, 153 (1976)
70. Fendler, E. J., et al.: J. Phys. Chem. *77*, 1432 (1973)
71. Fendler, J. H., et al.: J. Chem. Soc. Faraday Trans. *I 68*, 280 (1973)
72. Fendler, J. H., Fendler, E. J.: Micellar and macromolecular catalysis. New York: Academic 1975
73. Fendler, E. J., Constien, V. G., Fendler, J. H.: J. Phys. Chem. *79*, 917 (1975)
74. Fendler, J. H.: In: Micellization, solubilization and microemulsions. p. 695, Mittal, K. L. (ed.). New York: Plenum 1977
75. Flory, P. J.: Principles of polymer chemistry. Chap. 12, Ithaca, N. Y.: Cornell Univ. Press 1953
76. Ford Jr., N. C.: Chem. Scripta *2*, 193 (1972)
77. Frank, S. G., Zografi, G.: J. Pharm. Sci. *58*, 993 (1969)
78. Frisch, W. L., Simha, R.: In: Rheology, theory and applications. Vol. 1, p. 525, Eirich, F. (ed.) New York: Academic 1956
79. Friberg, S.: J. Am. Oil Chem. Soc. *48*, 578 (1971)
80. Friberg, S.: J. Colloid Interf. Sci. *29*, 155 (1969)
81. Fryar, A. J., Kaufman, S.: J. Colloid Interface Sci. *29*, 444 (1969)
82. Fowkes, F. M.: J. Phys. Chem. *66*, 1843 (1962)
83. Fowkes, F. M.: In: Solvent properties of surfactant solutions. Shinoda, K. (ed.). New York: Dekker 1967
84. Fuchs, O., Donle, H. L.: Trans. Faraday Soc. *30*, 707 (1934)
85. Fung, B. M., McAdams, J. L.: Biochim. Biophys. Acta *12*, 4928 (1974)
86. Gentile, F. P., Ricci, F., Podo, F.: Gaz. Chim. Ital. *106*, 423 (1976)
87. Gonick, E.: J. Colloid Sci. *1*, 343 (1946)
88. Goodwin, J. W.: In: Colloid Sci. Vol. 2, Chap. 7, London: The Chemical Soc. 1975
89. Heilweil, I. J.: J. Colloid Sci. *19*, 105 (1964)
90. Hildebrand, J. H., Scott, R. L.: The solubility of nonelectrolytes. New York: Dover 1964
91. Hildebrand, J. H., Prausnitz, J. M., Scott, R. L.: Regular and related solutions. New York: Van Nostrand Reinhold 1970
92. Honig, J. G., Singleterry, C. R.: J. Phys. Chem. *58*, 201 (1954)
93. Honig, J. G., Singleterry, C. R.: J. Phys. Chem. *60*, 1108 (1956)
94. Hopmann, R. F. W.: Ber. Bunsenges. Phys. Chem. *77*, 52 (1973)
95. Holzbach, R. T., et al.: In: Micellization, solubilization and microemulsions. Vol. 1, p. 403, Mittal, K. L. (ed.). New York: Plenum 1977
96. Huglin, M. B.: Determination of molecular weights by light scattering, Topics Curr. Chem. *77*, 141–232 (1978)
97. Hutchinson, E., Inaba, A., Bailey, L. G.: Z. Phys. Chem. *5*, 344 (1955)

98. Hutchinson, E.: J. Phys. Chem. *66*, 577 (1962)
99. Ito, K., Yamashita, Y.: J. Colloid Interface Sci. *19*, 152 (1964)
100. Jean, Y., Ache, H. J.: J. Am. Chem. Soc. (1978) in press
101. Jean, Y., Ache, H. J.: J. Am. Chem. Soc. *100*, 984 (1978)
102. IUPAC: Reporting experimental data dealing with critical micellization concentration (CMCs). Commision 1.6 on Colloid and Surface Chemistry. JUPAC Oxford 1976
103. Kaufman, S., Singleterry, C. R.: J. Colloid Sci. *10*, 139 (1955)
104. Kaufman, S., Singleterry, C. R.: J. Colloid Sci. *7*, 453 (1952)
105. Kaufman, S., Singleterry, C. R.: J. Colloid Sci. *12*, 469 (1957)
106. Kaufman, S., Singleterry, C. R.: J. Colloid Sci. *12*, 465 (1957)
107. Kaufman, S., Singleterry, C. R.: J. Phys. Chem. *62*, 1257 (1958)
108. Kerker, M.: Scattering of light and other electromagnetic radiation. New York: Academic 1969
109. Kertes, A. S., Levy, O., Markovits, G.: J. Phys. Chem. *74*, 3568 (1970)
110. Kertes, A. S., Markovits, G.: J. Phys. Chem. *72*, 4202 (1968)
111. Kertes, A. S., et al.: Proceedings VIth Intern. Conf. Surface Activity, Zürich 1972
112. Kertes, A. S., Gutmann, H.: Surf. Colloid. Sci. *8*, 193 (1975)
113. Kertes, A. S., Gutmann, H.: Surfactants in organic solvents. In: Surface and Colloid Science. Vol. 8, p. 193, Matijević, E. (ed.). New York: Wiley 1976
114. Kitahara, A., Ishikawa, T., Tanimori, S.: J. Colloid Interface Sci. *23*, 243 (1967)
115. Kitahara, A.: Bull. Chem. Soc. Japan *29*, 15 (1956)
116. Kitahara, A.: Bull. Chem. Soc. Japan *31*, 288 (1958)
117. Kitahara, A., Tobayashi, T., Tachibana, T.: J. Phys. Chem. *66*, 363 (1962)
118. Kitahara, A., Ishikawa, T.: J. Colloid Interface Sci. *24*, 189 (1967)
119. Kitahara, A.: In: Cationic surfactants. Jungermann, E. (ed.). New York: 1970
120. Kitahara, A., Kon-no, K.: Micelle formation in nonaqueous media. In: Colloidal dispersions and micellar behavior. ACS symposium series No. 9. 225, 1975
121. Kitahara, A.: Bull. Chem. Soc. Japan *28*, 234 (1955)
122. Klevens, H. B.: J. Phys. Colloid Chem. *51*, 1143 (1947)
123. Kon-no, K., Kitahara, A.: J. Colloid Interface Sci. *35*, 636 (1971)
124. Kon-no, K., Kitahara, A.: J. Colloid Interface Sci. *35*, 409 (1971)
125. Kon-no, K., Jin-no, T., Kitahara, A.: J. Colloid Interface Sci. *49*, 383 (1974)
126. Kon-no, K., Kitahara, A.: Kogyo Kagoku Zasshi *68*, 2058 (1965)
127. Koryta, J., Dvorak, J., Bohackova, V.: Electrochemistry, Wien: Springer 1975
128. Krafft, F., Wiglow, H.: Ber. dtsch. chem. Ges. *28*, 2566 (1895)
129. Kratohvil, J. P., Dellicolli, H. T.: Fed. Proc. *29*, 1335 (1970)
130. Kreutzer, J.: Z. Phys. Chem. B53, 213 (1943)
131. Kvita, P.: unpublished (to appear in J. Colloid Interface Sci.)
132. Levy, O., Markovits, G., Kertes, A. S.: J. Phys. Chem. *75*, 542 (1971)
133. Lissant, K. J. (ed.): Emulsions and emulsions technology. New York: Dekker 1974
134. Little, R. C., Singleterry, C. R.: J. Phys. Chem. *68*, 3453 (1964)
135. Little, R. C.: J. Phys. Chem. *74*, 1817 (1970)
136. Ljunggren, S., Lamm, O.: Acta Chem. Scand. *12*, 1834 (1958)
137. Lo, F. Y., et al.: J. Phys. Chem. *79*, 2609 (1975)
138. Madia, W. J., Nichols, A. L., Ache, H. J.: J. Am. Chem. Soc. *97*, 5041 (1975)
139. Magid, L.: Solvent effects on amphiphilic aggregation. In: Proceedings of the section on solution chemistry of surfactants held at the 52nd Colloid and Surface Science Sympos. Knoxville, TN 1978
140. Malik, W. U., Jain, K., Siddiqui, M. J.: Solutions of cobalt hexamine soaps in organic solvents. In: Proceedings Intern. Conf. Colloid and Surface Science. Vol. 1, p. 501, Wolfram, E. (ed,). Budapest: Akademiai Kiado 1975
141. Markovits, G., Kertes, A. S.: In: Solvent extraction chemistry. p. 390, Dryssen, D., Liljenzin, J. O., Rydberg, J. (eds.). Amsterdam: North Holland Publ. 1967
142. Mathews, M. B., Hirschhorn, E.: J. Colloid Sci. *8*, 86 (1953)
143. Mazer, N. A., Carey, M. C., Benedek, G. B.: In: Micellization, solubilization and microemulsions. Vol. 1, p. 359, Mittal, K. L. (ed.). New York: Plenum 1977

144. Mazer, N. A., Kwasnick, R. F., Benedek, G. B.: ibid, p. 383
145. Mazer, N. A., Benedek, G. B., Carey, M. C.: J. Phys. Chem. *80*, 1075 (1976)
146. Mehrotra, K. N., Mehtar, V. P., Nagar, T. N.: J. Prakt. Chem. *313*, 607 (1971)
147. Mehrotra, K. N.: Cellul. Chem. Technol. *7*, 387 (1973)
148. Merrigan, J. A., Tao, S. J., Green, J. H.: In: Physical methods of chemistry, Vol. I, part 3. Weissberger, D. A., Rossiter, B. W. (ed.). New York: Wiley 1972
149. Muckerjee, P.: Differing patterns of self-association and micelle formation: In: Physical chemistry: Enriching topics from colloid and surface science. Van Olphen, H., Mysels, K. J. (eds.). La Jolla, Calif. 1975
150. Muller, N.: Errors in micellization enthalpies from temperature dependence of critical micelle concentrations. In: Micellization, solubilization, and microemulsions. Vol. 1, p. 229, Mittal, K. L. (ed.). New York – London: Plenum 1977
151. Muller, N.: J. Colloid Interface Sci. *63*, 383 (1978)
152. Muller, N.: J. Phys. Chem. *79*, 287 (1975)
153. Muto, S., Meguro, K.: Bull. Chem. Soc. Japan *46*, 1316 (1973)
154. Muto, S., Shimazaki, Y., Meguro, K.: J. Colloid Interface Sci. *49*, 173 (1974)
155. Nakagawa, T., Shinoda, K.: Physicochemical studies in aqueous solutions of nonionic surface active agents. In: Colloidal surfactants. New York: Academic 1963
156. Nelson, S. M., Pink, R. C.: J. Chem. Soc. (London) *1952*, 1744
157. Nicholas, J. B., Ache, H. J.: J. Chem. Phys. *57*, 1599 (1977)
158. Oster, G.: In: Techniques of chemistry. Vol. 1, part III, Chap. 2, Weissberger, A. (ed.). New York: Wiley 1969
159. Ottewill, R. H., Walker, T.: Kolloid Z. Z. Polymere *227*, 108 (1968)
160. Palit, S. R., Venkateswarlu, V.: J. Chem. Soc. (London) *1954*, 2129
161. Peri, J. B.: J. Am. Oil Chem. Soc. *35*, 110 (1958)
162. Peri, J. B.: J. Colloid Interface Sci. *29*, 6 (1969)
163. Peri, J. B.: Paper presented before Colloid Div., 124th Meeting Am. Chem. Soc., Chicago, Sept. 6–11, 1953
164. Perrin, F.: J. Phys. Radium (VI) *7*, 390 (1962)
165. Perrin, F.: J. Phys. Radium *5*, 497 (1934)
166. Pilpel, N.: Trans. Faraday Soc. *56*, 893 (1960)
167. Pilpel, N.: Trans. Faraday Soc. *57*, 1426 (1961)
168. Poland, D.: Cooperative equilibria in physical biochemistry, Oxford: Oxford University Press 1978
169. Preston, W. C.: J. Phys. Colloid Chem. *52*, 84 (1948)
170. Prigogine, I., Defay, R.: Chemical thermodynamics. London – New York – Toronto: Longmans Green & Co. 1954
171. Prince, L. M.: J. Colloid Interface Sci. *52*, 182 (1975)
172. Prince, L. M.: J. Colloid Interface Sci. *29*, 216 (1969)
173. Ray, A.: J. Am. Chem. Soc. *11*, 6511 (1969)
174. Ray, A.: Nature *231*, 313 (1971)
175. Reerink, H.: J. Colloid Sci. *20*, 217 (1965)
176. Reerink, H.: J. Colloid Sci. *20*, 257 (1965)
177. Reerink, H.: Proc. 3rd Intern. Congr. Surface Activity 1960, Vol. 1, Section A
178. Robinson, N.: J. Pharm. Pharmacol. *12*, 685 (1960)
179. Rothrock, D. A., Kraus, C. A.: J. Am. Chem. Soc. *59*, 1699 (1973)
180. Rutgers, R.: Rheol. Acta *4*, 305 (1962)
181. Ruckenstein, E., Chi, J.: J. Chem. Soc. Faraday Trans. II, *71*, 1690 (1975)
182. Saito, H., Shinoda, K.: J. Colloid Interface Sci. *32*, 647 (1970)
183. Saito, H., Shinoda, K.: J. Colloid Interface Sci. *35*, 359 (1971)
184. Schick, M. J. (ed.): Nonionic surfactants. New York: Dekker 1974
185. Scriven, L. E.: Discussion remarks to contribution of Eicke, H. F. in Micellization, solubilization and microemulsions. Mittal, K. L. (ed.). New York: Plenum 1977
186. Shah, D. O.: 48th Natl. Colloid Symposium Austin, Texas, June 1974, preprints p. 173

187. Shaw, C. M., Johnson, J. F.: In: Offic. Dig. Federation Paint and Varnish Prod. Clubs No. 339: 216 (April 1953)
188. Sheffer, H.: Can. J. Res. *26B*, 481 (1948)
189. Sherman, P. (ed.): Emulsion science. New York: Academic 1968
190. Shinoda, K., Nakagawa, T., Tamamushi, B., and Isemura, T.: Colloidal surfactants. New York – London: Academic 1963
191. Shinoda, K.: Bull. Chem. Soc. Japan *26*, 101 (1953)
192. Shinoda, K., Hutchinson, E.: J. Phys. Chem. *66*, 577 (1962)
193. Shinoda, K., Ogawa, T.: J. Colloid Interface Sci. *24*, 56 (1967)
194. Shinoda, K., Kunieda, H.: J. Colloid Interface Sci. *42*, 381 (1973)
195. Shinoda, K., Friberg, S.: Adv. Colloid Interface Sci. *4*, 281 (1975)
196. Singleterry, C. R., Weinberger, L. A.: J. Am. Chem. Soc. *73*, 4574 (1951)
197. Singleterry, C. R.: J. Am. Oil Chem. Soc. *32*, 446 (1955)
198. Sirianni, A. F., Cowie, J. M. G., Puddington, I., E.: Can. J. Chem. *40*, 957 (1962)
199. Sirianni, A. F., Coleman, R. D.: Can. J. Chem. *42*, 682 (1964)
200. Spegt, P., Skoulios, A.: J. Chim. Phys. *62*, 377 (1965)
201. Soldatov, V., Oedberg, L., Högfeldt, E.: Infrared and NMR studies of lithium and cesium salts of dinonylnaphthalene sulfonic acid in heptane. In: Ion exchange and membranes. Vol. 2, p. 83, Gordon and Breach 1975
202. Stauff, J.: Kolloidchemie. Berlin – Göttingen – Heidelberg: Springer 1960
203. Strahm, U., Ph D Thesis, Univ. Basel 1977
204. Strehlow, H., Becker, M.: Z. Elektrochem. Ber. Bunsenges. Phys. Chem. *63*, 457 (1959)
205. Tanford, C.: Physical chemistry of macromolecules. New York: Wiley 1961
206. Tanford, C.: The hydrophilic effect. New York: Wiley 1973
207. Tavernier, S.: Inverted micellar structure of anioniç surfactants in apolar medium. Proefschrift. Universiteit Antwerpen 1977
208. Tokiwa, F., Isemura, T.: Bull. Chem. Soc. Japan *35*, 1737 (1962)
209. Ueno, Masaharn: Bull. Chem. Soc. Japan *49*, 1776 (1976)
210. Ueno, M., Kishimoto, H.: Bull. Chem. Soc. Japan *50*, 1637 (1977)
211. Ueno, M., Kishimoto, H., Kyogoku, Y.: J. Colloid Interface Sci. *63*, 113 (1978)
212. Vasta, G., Diplom Thesis, Univ. Basel 1979
213. van der Waarden, M.: J. Colloid Sci. *5*, 448 (1950)
214. Wagner, C.: Colloid & Polymer Sci. 254, 400 (1976)
215. Walter, W. V., Hayes, F. B.: Biochim. Biophys. Acta *249*, 528 (1971)
216. Ware, W. R.: In: Creation and detection of the excited state. Lamda, A. (ed.). New York: Dekker 1971
217. Weissberger, D. A. (ed.). New York: Wiley 1972, Vol. 1, Part III
218. Williams, E., et al.: J. Am. Chem. Soc. *95*, 4871 (1973)
219. Williams, T. L., Ache, H. J.: J. Chem. Phys. *50*, 4493 (1969)
220. Winkelmair, D.: Arch. Biochem. Biophys. *147*, 509 (1971)
221. Wong, M., Thomas, J. K., Grätzel, M.: J. Am. Chem. Soc. *98*, 2391 (1976)
222. Wong, M., Thomas, J. K., Nowak, T.: J. Am. Chem. Soc. *99*, 4730 (1977)
223. Wong, M., Thomas, J. K.: In: Micellization, solubilization, and microemulsions Vol. 2, p. 647, Mittal, K. L. (ed.). New York: Plenum 1977
224. Zana, R.: Ultrasonic absorption studies of solutions of ionic amphiphiles in organic solvents. In: Proc. section on solution chem. surfactants, 52nd colloid and surface sci. symposium. Knoxville 1978
225. Zimm, B. H.: J. Chem. Phys. *16*, 1099 (1948)
226. Zinsli, P. W., Eicke, H. F.: Progr. Colloid Polymer Sci. *65*, 158 (1978)
227. Zulauf, M., Eicke, H. F.: J. Phys. Chem. *83*, 480 (1979)
228. Zundel, G.: Hydration and intermolecular interaction. New York: Academic 1969

Received April 9, 1979

Author Index Volumes 26–87

The volume numbers are printed in italics

Albini, A., and Kisch, H.: Complexation and Activation of Diazenes and Diazo Compounds by Transition Metals. *65*, 105–145 (1976).

Altona, C., and Faber, D. H.: Empirical Force Field Calculations. A Tool in Structural Organic Chemistry. *45*, 1–38 (1974).

Anderson, D. R., see Koch, T. H.: *75*, 65–95 (1978).

Anderson, J. E.: Chair-Chair Interconversion of Six-Membered Rings. *45*, 139–167 (1974).

Anet, F. A. L.: Dynamics of Eight-Membered Rings in Cyclooctane Class. *45*, 169–220 (1974).

Ariëns, E. J., and Simonis, A.-M.: Design of Bioactive Compounds. *52*, 1–61 (1974).

Ashfold, M. N. R., Macpherson, M. T., and Simons, J. P.: Photochemistry and Spectroscopy of Simple Polyatomic Molecules in the Vacuum Ultraviolet. *86*, X–X (1979).

Aurich, H. G., and Weiss, W.: Formation and Reactions of Aminyloxides. *59*, 65–111 (1975).

Balzani, V., Bolletta, F., Gandolfi, M. T., and Maestri, M.: Bimolecular Electron Transfer Reactions of the Excited States of Transition Metal Complexes. *75*, 1–64 (1978).

Bardos, T. J.: Antimetabolites: Molecular Design and Mode of Action. *52*, 63–98 (1974).

Barnes, D. S., see Pettit, L. D.: *28*, 85–139 (1972).

Bauder, A., see Frei, H.: *81*, 1–98 (1979).

Bastiansen, O., Kveseth, K., and Møllendal, H.: Structure of Molecules with Large Amplitude Motion as Determined from Electron-Diffraction Studies in the Gas Phase. *81*, 99–172 (1979).

Bauer, S. H., and Yokozeki, A.: The Geometric and Dynamic Structures of Fluorocarbons and Related Compounds. *53*, 71–119 (1974).

Baumgärtner, F., and Wiles, D. R.: Radiochemical Transformations and Rearrangements in Organometallic Compounds. *32*, 63–108 (1972).

Bayer, G., see Wiedemann, H. G.: *77*, 67–140 (1978).

Bernardi, F., see Epiotis, N. D.: *70*, 1–242 (1977).

Bernauer, K.: Diastereoisomerism and Diastereoselectivity in Metal Complexes. *65*, 1–35 (1976).

Bikerman, J. J.: Surface Energy of Solids. *77*, 1–66 (1978).

Boettcher, R. J., see Mislow, K.: *47*, 1–22 (1974).

Bolletta, F., see Balzani, V.: *75*, 1–64 (1978).

Brandmüller, J., and Schrötter, H. W.: Laser Raman Spectroscopy of the Solid State. *36*, 85–127 (1973).

Bremser, W.: X-Ray Photoelectron Spectroscopy. *36*, 1–37 (1973).

Breuer, H.-D., see Winnewisser, G.: *44*, 1–81 (1974).

Brewster, J. H.: On the Helicity of Variously Twisted Chains of Atoms. *47*, 29–71 (1974).

Brocas, J.: Some Formal Properties of the Kinetics of Pentacoordinate Stereoisomerizations. *32*, 43–61 (1972).

Brown, H. C.: Meerwein and Equilibrating Carbocations. *80*, 1–18 (1979).

Brunner, H.: Stereochemistry of the Reactions of Optically Active Organometallic Transition Metal Compounds. *56*, 67–90 (1975).

Buchs, A., see Delfino, A. B.: *39*, 109–137 (1973).

Bürger, H., and Eujen, R.: Low-Valent Silicon. *50*, 1–41 (1974).

Burgermeister, W., and Winkler-Oswatitsch, R.: Complexformation of Monovalent Cations with Biofunctional Ligands. *69*, 91–196 (1977).

Schutte, C. J. H.: The Infra-Red Spectra of Crystalline Solids. *36*, 57–84 (1973).

Schwarz, H.: Some Newer Aspects of Mass Spectrometric *Ortho* Effects. *73*, 231–263 (1978).

Schwedt, G.: Chromatography in Inorganic Trace Analysis. *85*, 159–212 (1979).

Scrocco, E., and Tomasi, J.: The Electrostatic Molecular Potential as a Tool for the Interpretation of Molecular Properties. *42*, 95–170 (1973).

Sears, P. G., see Lemire, R. J.: *74*, 45–91 (1978).

Shaik, S., see Epiotis, N. D.: *70*, 1–242 (1977).

Sharp, K. G., see Margrave, J. L.: *26*, 1–35 (1972).

Sheldrick, W. S.: Stereochemistry of Penta- and Hexacoordinate Phosphorus Derivatives. *73*, 1–48 (1978).

Shue, H.-J., see Gelernter, H.: *41*, 113–150 (1973).

Simonetta, M.: Qualitative and Semiquantitative Evaluation of Reaction Paths. *42*, 1–47 (1973).

Simonis, A.-M., see Ariëns, E. J.: *52*, 1–61 (1974).

Simons, J. P., see Ashfold, M. N. R.: *86*, X–X (1979).

Sluski, R. J., see Koch, T. H.: *75*, 65–95 (1978).

Smith, S. L.: Solvent Effects and NMR Coupling Constants. *27*, 117–187 (1972).

Sørensen, G. O.: New Approach to the Hamiltonian of Nonrigid Molecules. *82*, 97–175 (1979).

Spanget-Larsen, J., see Gleiter, R.: *86*, X–X (1979).

Špirko, V., see Papousek, D.: *68*, 59–102 (1976).

Sridharan, N. S., see Gelernter, H.: *41*, 113–150 (1973).

Stohrer, W.-D., Jacobs, P., Kaiser, K. H., Wich, G., and Quinkert, G.: Das sonderbare Verhalten electronen-angeregter 4-Ringe-Ketone. – The Peculiar Behavior of Electronically Exited 4-Membered Ring Ketones. *46*, 181–236 (1974).

Stoklosa, H. J., see Wasson, J. R.: *35*, 65–129 (1973).

Suhr, H.: Synthesis of Organic Compounds in Glow and Corona Discharges. *36*, 39–56 (1973).

Sutter, D. H., and Flygare, W. H.: The Molecular Zeeman Effect. *63*, 89–196 (1976).

Tacke, R., and Wannagat, U.: Syntheses and Properties of Bioactive Organo-Silicon Compounds. *84*, 1–75 (1979).

Thakkar, A. J.: The Coming of the Computer Age to Organic Chemistry. Recent Approaches to Systematic Synthesis Analysis. *39*, 3–18 (1973).

Tölg, G., see Rüssel, H.: *33*, 1–74 (1972).

Tomasi, J., see Scrocco, E.: *42*, 95–170 (1973).

Trinjastič, N., see Gutman, I.: *42*, 49–93 (1973).

Trost, B. M.: Sulfuranes in Organic Reactions and Synthesis. *41*, 1–29 (1973).

Tsigdinos, G. A.: Heteropoly Compounds of Molybdenum and Tungsten. *76*, 1–64 (1978).

Tsigdinos, G. A.: Sulfur Compounds of Molybdenum and Tungsten. Their Preparation, Structure, and Properties. *76*, 65–105 (1978).

Tsuji, J.: Organic Synthesis by Means of Transition Metal Complexes: Some General Patterns. *28*, 41–84 (1972).

Turley, P. C., see Wasserman, H. H.: *47*, 73–156 (1974).

Ugi, I., see Dugundji, J.: *39*, 19–64 (1973).

Ugi, I., see Dugundji, J.: *75*, 165–180 (1978).

Ugi, I., see Gasteiger, J.: *48*, 1–37 (1974).

Ullrich, V.: Cytochrome P450 and Biological Hydroxylation Reactions. *83*, 67–104 (1979).

Veal, D. C.: Computer Techniques for Retrieval of Information from the Chemical Literature. *39*, 65–89 (1973).

Vennesland, B.: Stereospecifity in Biology. *48*, 39–65 (1974).

Veprek, S.: A Theoretical Approach to Heterogeneous Reactions in Non-Isothermal Low Pressure Plasma. *56*, 139–159 (1975).

Vilkov, L., and Khaikin, L. S.: Stereochemistry of Compounds Containing Bonds Between Si, P, S, Cl, and N or O. *53*, 25–70 (1974).

J. J. Bikerman

Foams

1973. 79 figures. IX, 337 pages
(Applied Physics and Engineering, Volume 10)
ISBN 3-540-06108-8

Contents: General. Foam Films. – Formation and Structure. – Measurement of Foaminess. – Results of Foaminess Measurements. – Three-phase Foams. – Foam Drainage. – Mechanical Properties of Foams. – Optical Properties of Foams. – Electric Properties of Foams. – Theories of Foam Stability. – Foams in Nature and Industry. – Applications. Separation by Foam. – Other Applications.

"... The great merits of Bikerman's work are the compilation of an extensive experimental material, the exceptionally concise and clear style, the demonstration of the subject on hand, of simple numerical examples, and the richness in experimental and methodological detail. Therefore, notwithstanding a certain one-sidedness, the book is a valuable contribution to the literature on surface chemistry."
Acta Chimica

"... this book can be read on two levels, either selectively by a student new to the subject or extensively by an established researcher interested in a thorough and authoritative review."
J. of the American Chemical Soc.

Springer-Verlag
Berlin
Heidelberg
New York

Topics in Current Chemistry

Fortschritte der chemischen Forschung
Managing Editor: F. L. Boschke

A Selection

Springer-Verlag
Berlin Heidelberg New York